AF667625

Synthesis Lectures on Engineering, Science, and Technology

The focus of this series is general topics, and applications about, and for, engineers and scientists on a wide array of applications, methods and advances. Most titles cover subjects such as professional development, education, and study skills, as well as basic introductory undergraduate material and other topics appropriate for a broader and less technical audience.

Mourad Talbi · Brahim Nasraoui

Image Denoising Using Mathematical Tools

Springer

Mourad Talbi
LMEEVED Laboratory
Center of Researches and Technologies
of Energy of Borj Cedria
Ben Arous, Tunisia

Brahim Nasraoui
Department of Computer Sciences, University
College of Duba
University of Tabuk
Duba, Saudi Arabia

ISSN 2690-0300 ISSN 2690-0327 (electronic)
Synthesis Lectures on Engineering, Science, and Technology
ISBN 978-3-032-11877-6 ISBN 978-3-032-11878-3 (eBook)
https://doi.org/10.1007/978-3-032-11878-3

© The Editor(s) (if applicable) and The Author(s), under exclusive license to Springer
Nature Switzerland AG 2026

This work is subject to copyright. All rights are solely and exclusively licensed by the Publisher, whether the whole or part of the material is concerned, specifically the rights of translation, reprinting, reuse of illustrations, recitation, broadcasting, reproduction on microfilms or in any other physical way, and transmission or information storage and retrieval, electronic adaptation, computer software, or by similar or dissimilar methodology now known or hereafter developed.
The use of general descriptive names, registered names, trademarks, service marks, etc. in this publication does not imply, even in the absence of a specific statement, that such names are exempt from the relevant protective laws and regulations and therefore free for general use.
The publisher, the authors and the editors are safe to assume that the advice and information in this book are believed to be true and accurate at the date of publication. Neither the publisher nor the authors or the editors give a warranty, expressed or implied, with respect to the material contained herein or for any errors or omissions that may have been made. The publisher remains neutral with regard to jurisdictional claims in published maps and institutional affiliations.

This Springer imprint is published by the registered company Springer Nature Switzerland AG
The registered company address is: Gewerbestrasse 11, 6330 Cham, Switzerland

If disposing of this product, please recycle the paper.

Introduction

This book is intended for students, engineers, and researchers interested in the field of image processing. It introduces and analyzes a number of image denoising techniques existing in literature. It contains three chapters where in the first, the authors detail image denoising approaches employing the Adaptive Frequency Median Filter (AFMF). In the second chapter, the authors present their image denoising technique based on two different wavelet transforms, which are 2-D dual-tree DWT and Stationary Wavelet Transform 2-D. Finally, in the third chapter, the authors discuss their technique of MRI Denoising based on 2-D Dual-Tree DWT and Stationary Wavelet Transform 2-D.

Contents

List of Figures

Image Denoising Approaches Employing Adaptive Frequency Median Filter (AFMF)

1

1.1 Introduction

Image denoising [1–3] is an interesting task in the pre-processing step in order to ameliorate the image quality for further post-processing tasks. Among these tasks, we can mention: image analysis, feature extraction, and image segmentation. Due to its important role, it has attracted much attention [4–8]. The aim of the image denoising task is to cancel noise by preserving edges, structures, textures, and other image details [9–11]. There are many kinds of noise, including the Gaussian noise [12], the Poisson noise [13–15], and the impulse noise [16–20]. The salt and pepper noise is a simple type of impulse noise and it is common in many electronic devices. The noisy image involving the salt and pepper noise can be formulated as follows [21–25]:

$$y_{\mathrm{ij}} = \begin{cases} \delta_{\min}, \text{ with probability} p \\ \delta_{\max}, \text{ with probability} q \\ x_{\mathrm{ij}}, \text{ with probability} 1 - (p + q), \end{cases} \tag{1.1}$$

where x_{ij} and y_{ij} are respectively the clean image and the corresponding noisy image at the pixel location, (i, j). The values $\delta_{\max}$ and $\delta_{\min}$ are respectively the maximum and minimum grey values of the image. The probabilities are $1 \geq q + p, q, p \geq 0$. The sum $q + p$ is a noise level (noise density, noise ratio) of the salt and pepper noise. Uğur Erkan et al. [7] have considered $p = q$ as the case where the noise level is even. Many approaches were introduced for images denoising, including regularizeation, filtering, wavelet, mathematical transformations, machine learning, and statistical methods [8]. Respecting the filtering method, the median filters are very effective for eliminating the salt and pepper noise [8]. Among those filters, we can mention the median filter (MF) and its adaptive versions [26]. The MF implementation is easy, which is starting from a window (owning a

© The Author(s), under exclusive license to Springer Nature Switzerland AG 2026

M. Talbi and B. Nasraoui, *Image Denoising Using Mathematical Tools*, Synthesis Lectures on Engineering, Science, and Technology,
https://doi.org/10.1007/978-3-032-11878-3_1

fixed size) inside the image. At each step, the median value of the pixels in the considered window is evaluated and assigned to the centre pixel of the window [10]. Diverse sizes of the window can be used in MF, such as 3 × 3, 5 × 5, and so on. A window having a larger size is providing a high blur effect and a window owning a small size is providing a less blur effect [27]. The MF is an effective filter for eliminating the low density salt and pepper noise [28]. An MF limit is effective in cancelling the low density salt and pepper noise [8]. It doesn't work well in the case of the high density salt and pepper noise. When processing the high density noise, it is essential to use larger windows. Though, MF can't preserve well edges and textures. In order to discard this drawback, centre CMF [29] and weighted MF (WMF) [30, 31] were proposed. Despite these filters are better than MF, they are still not effective for cancelling the high density salt and pepper noise. In order to ameliorate the performance of the median-based filters, a dynamic window has been applied [32]. The application of a dynamic window was found to be effective in order to process a high density noise. The size of the window starts from 3 × 3 and enlarges gradually until an adaptive condition is fulfilled or a limited value is reached by this size [22, 33]. The adaptive condition may vary depending on each adaptive filter. In [3], we will detail an image denoising technique which combining the self-organizing migration algorithm (SOMA) [34] and the adaptive frequency median filter (AFMF) [8]. This technique consists firstly in applying an adaptive frequency median filter (AFMF) to the noisy image for obtaining the first version of the denoised image. The latter is considered as the clean image and employed as an input of an image denoising system applying SOMA [34]. The latter is then applied for denoising the noisy image and having a final version of the denoised image. The remaining of this chapter is arranged as follows: In part 1.2, the AFMF process is detailed; the part 1.3 is devoted to SOMA, and in part 1.4, we will detail our image denoising technique introduced in [3]. In part 1.5, results and discussion will be presented, and in part 1.6, we will conclude.

1.2 Materials and Techniques

Adaptive Frequency Median Filter (AFMF)

The matrices $A := \left[a_{\mathrm{ij}}\right]_{m\times n}$ and $B := [b_{\mathrm{kl}}]_{m\times n}$ are two images so that $m \geq 9$ and $n \geq 9$; for all i and j and b_{ij} and a_{ij} designate integer numbers belonging to [0, 255].

Definition 1: Let $A := \left[a_{\mathrm{ij}}\right]_{m,n}$ and $t \in \{1, 2, \ldots, \min\{m, n\}\}$ [8]. The matrix $\left[\overline{a}_{\mathrm{rs}}\right]_{(m+2t)\times(n+2t)}$ designates the t-symmetric pad matrix of A is denoted $\underset{t-\mathrm{sym}}{\overset{=}{A}}$ (or $\underset{t}{\overset{=}{A}}$).

Example 1: Let $A = \begin{bmatrix} 11 & 12 & 13 \\ 21 & 22 & 255 \\ 0 & 32 & 33 \end{bmatrix}$ [8]. Then,

$$\underset{3}{\overset{=}{A}} = \begin{bmatrix} 33 & 32 & 0 & 0 & 32 & 33 & 33 & 32 & 0 \\ 255 & 22 & 21 & 21 & 22 & 255 & 255 & 22 & 21 \\ 13 & 12 & 11 & 11 & 12 & 13 & 13 & 12 & 11 \\ 13 & 12 & 11 & 11 & 12 & 13 & 13 & 12 & 11 \\ 255 & 22 & 21 & 21 & 22 & 255 & 255 & 22 & 21 \\ 33 & 32 & 0 & 0 & 32 & 33 & 33 & 32 & 0 \\ 33 & 32 & 0 & 0 & 32 & 33 & 33 & 32 & 0 \\ 255 & 22 & 21 & 21 & 22 & 255 & 255 & 22 & 21 \\ 13 & 12 & 11 & 11 & 12 & 13 & 13 & 12 & 11 \end{bmatrix}_{9\times 9}$$

Definition 2: Let $A := \left[a_{\text{ij}}\right]_{m\times n}$ and $k \in \{1, 2, \ldots, t\}$ [8]. Then, the matrix A_{ij}^{k} owns the following expression:

$$\begin{bmatrix} \underset{(i+t-k)(j+t-k)}{\overset{=}{a}} & \cdots & \underset{(i+t-k)(j+t+k)}{\overset{=}{a}} \\ \vdots & \underset{(i+t)(j+t)}{\overset{=}{a}} & \vdots \\ \underset{(i+t+k)(j+t-k)}{\overset{=}{a}} & \cdots & \underset{(i+t+k)(j+t+k)}{\overset{=}{a}} \end{bmatrix}_{(2k+1)\times(2k+1)}$$

is named k-approximate matrix of a_{ij} in the matrix, $\underset{t}{\overset{=}{A}}$.

Example 2: When considering the Example 1, then we have:

$$A_{21}^{1} = \begin{bmatrix} \underset{43}{\overset{=}{a}} & \underset{44}{\overset{=}{a}} & \underset{45}{\overset{=}{a}} \\ \underset{53}{\overset{=}{a}} & \underset{54}{\overset{=}{a}} & \underset{55}{\overset{=}{a}} \\ \underset{63}{\overset{=}{a}} & \underset{64}{\overset{=}{a}} & \underset{65}{\overset{=}{a}} \end{bmatrix} = \begin{bmatrix} 11 & 11 & 12 \\ 21 & 21 & 22 \\ 0 & 0 & 32 \end{bmatrix}$$

Definition 3: Let the matrix, $A := \left[a_{\text{ij}}\right]_{m\times n}$ [8], then the matrix $\tilde{A} := [\tilde{a}_{1u}]_{1\times(m\cdot n)}$ is consisting of all entries (elements) of A and, due to being non-decreasing, it is called entry matrix (EM) of A.

Example 3: Let us considering Example 2 [8], then $\tilde{A}_{21}^{1} = [0011111221212232]$.

Definition 4: Let $A := \left[a_{\text{ij}}\right]_{m \times n}$ and $B := [b_{\text{kl}}]_{m \times n}$, then $\left\{b_{\text{kl}} : a_{\text{ij}} = b_{\text{kl}}\right\}$ designates the equivalence class of a_{ij} according to B and is denoted by $\left(a_{\text{ij}}\right)_B$. In case where $A = B$, the matrix, $\left(a_{\text{ij}}\right)$ can be used instead of $\left(a_{\text{ij}}\right)_B$.

Example 4: Let $A = \begin{bmatrix} 161 & 168 & 155 \\ 165 & 156 & 0 \\ 0 & 142 & 0 \end{bmatrix}$ and $B = \begin{bmatrix} 160 & 162 & 0 \\ 161 & 163 & 160 \\ 161 & 0 & 165 \end{bmatrix}$, then $(a_{23})_B = \{b_{13}, b_{32}\}$, $|(a_{23})_B| = 2$, $(a_{23}) = \{a_{23}, a_{31}, a_{33}\}$, and $|(a_{23})| = 3$.

Definition 5: Let the matrix, $\boldsymbol{A} := \left[\boldsymbol{a}_{\mathbf{ij}}\right]_{\boldsymbol{m \times n}}$, and $\boldsymbol{H}$ is the set of entries of the matrix, $\boldsymbol{A}$ and $|\boldsymbol{H}|$ is the $\boldsymbol{H}$ cardinality. The matrix, $\vec{A} := \left[\vec{a}_{1V}\right]_{1 \times |H|}$ is constituting of all entries of $\boldsymbol{A}$, and being strictly increasing, it is called a strictly increasing EM (SI-EM) of $\boldsymbol{A}$.

Definition 6: Let the matrix, $\vec{A} := \left[\vec{a}_{1V}\right]_{1 \times |H|}$ designates a SIEM of the matrix, $A := \left[a_{\text{ij}}\right]_{m \times n}$. The following value can be obtained:

$$\mathbf{fmed}\left(\vec{A}\right) := \begin{cases} \vec{\boldsymbol{a}}_{1((|\boldsymbol{H}|+1)/2)}, \ |\boldsymbol{H}| \text{is odd} \\ \vec{\boldsymbol{a}}_{1((|\boldsymbol{H}|+2)/2)}, \ |\boldsymbol{H}| \text{is even and} \left|\left(\vec{\boldsymbol{a}}_{1(|\boldsymbol{H}|/2)}\right)_{\tilde{A}}\right| \leq \left|\left(\vec{\boldsymbol{a}}_{1((|\boldsymbol{H}|+2)/2)}\right)_{\tilde{A}}\right| \\ \vec{\boldsymbol{a}}_{1(|\boldsymbol{H}|/2)}, \ \text{otherwise} \end{cases} \quad (1.2)$$

A frequency median of $\vec{A}$ with $\left|\left(\vec{a}_{1(|H|/2)}\right)_{\tilde{A}}\right|$ is the frequency of $\vec{a}_{1(|H|/2)}$ in $\tilde{A}$.

A frequency median of $\vec{A}$ with $\left|\left(\vec{a}_{1(|H|/2)}\right)_{\tilde{A}}\right|$ is the frequency of $\vec{a}_{1(|H|/2)}$ in $\tilde{A}$.

The fmed (frequency median) is working on a SIEM. In [8], all pixels are considered in SIEM, and fmed is evaluated such as the standard median when the pixel number was odd. In case where the pixel number is even [8], only the pair of pixels centre is considered and the fmed is the most repeated pixel value from the pair. In case where two pixels centre own the same number of repetitions, then fmed is the first pixel value from the right side [8]. The fmed is called frequency median due to the number of repetitions, which can be viewed as a frequency [8]. For having an outlook of AMF [8], the AMF detail is given as the Algorithm 1 (Fig. 1.1). The AFMF introduced in an earlier work [8] is presented as the Algorithm 2 (Fig. 1.2).

In [8], a flowchart was provided for evaluating the frequency median, fmed $(\vec{A}_{\text{ij}}^{k})$, as illustrated at Fig. 1.3. Here, for simplifying the flowchart, the following notions are given:

Compute $\bar{\bar{A}}_{k_{max}}$ for $A := [a_{ij}]_{m\times n}, k_{max} = 9$

For every pixel (i, j) of the image,

  For k from 1 to k_{max}

    If $(\min A_{ij}^{k} < med\ A_{ij}^{k})$ **AND** $(med\ A_{ij}^{k} < max\ A_{ij}^{k})$

        AND $(a_{ij} = min\ A_{ij}^{k}$ **OR** $a_{ij} = max\ A_{ij}^{k})$

      Set $a_{ij} \leftarrow med\ A_{ij}^{k}$

      Break

    End

  End

End

Fig. 1.1 Algorithm 1: AMF [8]

- $\text{sort}^{+}(\cdot)$: sorts values through the strictly increasing order,
- $|\text{SIEM}|$: cardinaity of SIEM,
- $|s_i|$: frequency (number of duplicates) of grey value s_i,
- $lef(\bullet)$: takes the first left value of the list,
- $righ(\bullet)$: takes the first right value of the list,
- mid $(\bullet)$: takes the middle values of the list. The mid function is returning one value in case where the number of elements of the list is odd, and it is returning two values in case where the number of elements of the list is even.

1.3 Self-organizing Migration Algorithm (SOMA)

Zijian Cao et al. [34] have introduced the self-organizing migration algorithm (SOMA). As a sort of swarm intelligence optimization algorithm, SOMA was introduced by Zelinka [35]. Like the ant colony algorithm (ACO) and particle swarm optimization (PSO) algorithm, SOMA is belonging to the class of cluster intelligence [34]. In this algorithm, at first, the individuals are initialized to form the initialization population and own the fitness value corresponding to each particle according to the specific problem to be solved

Compute $\bar{\bar{A}}_{k_{max}}$ for $A := [a_{ij}]_{m\times n}, k_{max} = 9$

For every pixel (i, j) of the image,

  For k from 1 to k_{max}

    If $(\min A_{ij}^k < med\ A_{ij}^k)$ **AND** $(med\ A_{ij}^k < max\ A_{ij}^k)$

      AND $(a_{ij} = min\ A_{ij}^k$ **OR** $a_{ij} = max\ A_{ij}^k)$

      Evaluate $\tilde{A}_{ij}^k$

      Evaluate $\vec{A}_{ij}^k$

      Set $a_{ij} \leftarrow \text{fmed}(\vec{A}_{ij}^k)$

      Break

    End

  End

Fig. 1.2 Algorithm 2: AFMF [8]

[34]. Other individuals in the population except the leader are migrated and updated with a migration loop (ML) [34]. The leader is the individual with the best fitness value in the current migration loop, and particles use a certain step for making a small jump to the leader in each migration loop [34]. A perturbation vector, PRT vector, is randomly initialized for guiding individuals move towards the leader, where the PRT vector is adjusted by setting a coefficient PRT, which often taking a value in the range, [0,1]. The way of individual migration is carried out according to the following equation:

$$X_{\text{temp}}^{\text{ML}} = X_{i,\text{state}}^{\text{ML}} + (X_L^{\text{ML}} - X_{i,\text{state}}^{\text{ML}}) \cdot t \cdot \text{PRTVector}. \tag{1.3}$$

In [8], $X_{i,\text{state}}^{\text{ML}}$ is representing the individuals to be migrated, X_L^{ML} is representing the leader in the ML generation migration loop, and $X_{\text{temp}}^{\text{ML}}$ is representing the new position obtained by individuals' migration and t designates the step length [8].

The individual migration in the population doesn't continue all the time. In case where the accumulation step attains the path length (maximum migration length), the migration loop is ceased. Then, the newest leader is selected for the next loop of migration, and finally, the algorithm terminates when the stop condition is met. The basic SOMA can be summarized by the following points [34]:

- NP individuals are initialized as the first generation migration population.

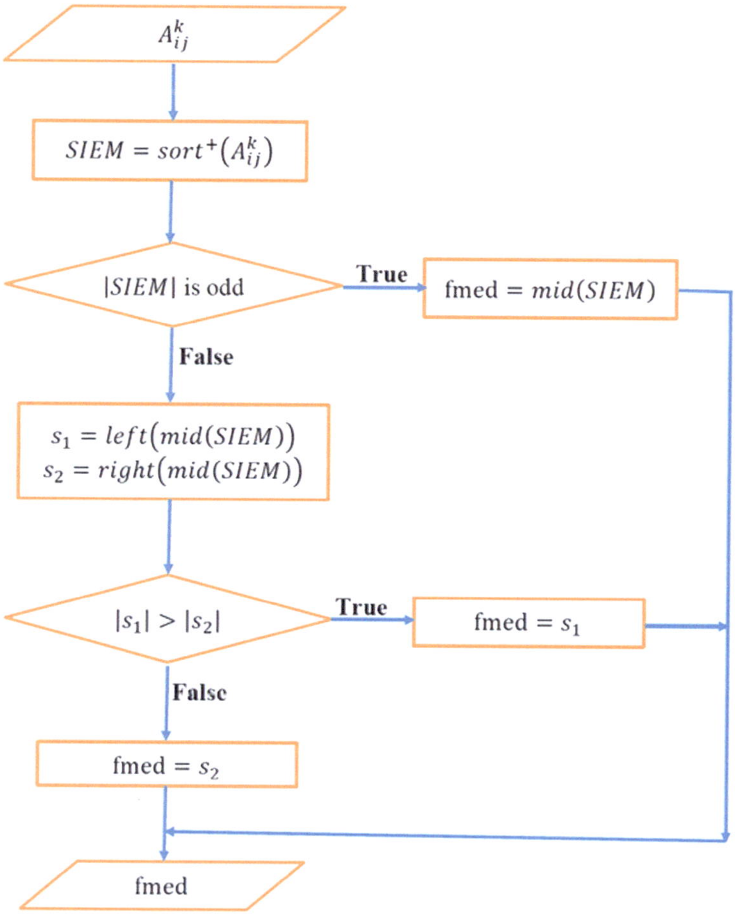

Fig. 1.3 Flowchart for evaluating the frequency median [8]

- The fitness value of each one of individuals is calculated, and the optimal individual is selected as the leader to guide other individuals to migrate.
- Individuals other than the leader are migrated according to Eq. (1.3), and the PRT vector value is updated.
- In case where the stop condition is attained, the migration stops and the optimal individual value will be returned.

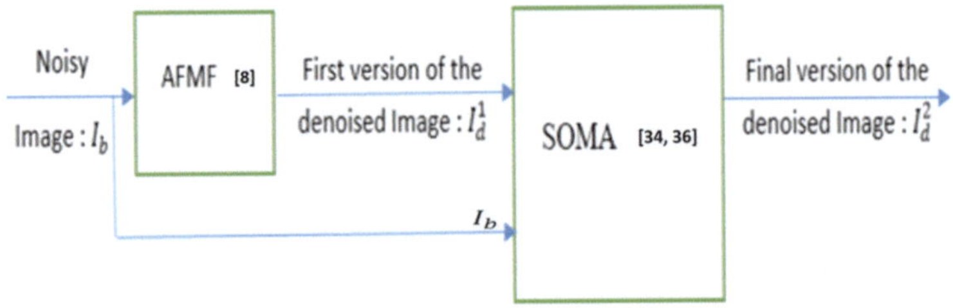

Fig. 1.4 Block diagram of the first image denoising technique introduced in [3]

1.4 The Proposed Image Denoising Technique

In this chapter, we detail an image denoising technique proposed in [3]. This technique is a hybrid approach combining the self-organizing migration algorithm (SOMA) [34, 36] and the adaptive frequency median filter (AFMF). The block diagram of this proposed denoising approach is illustrated in Fig. 1.4. According to the latter, the first step of this technique consists of applying AFMF [8] to the noisy image, I_b, to obtain the first version of the denoised image, denoted as I_d^1. The latter is considered as the clean image, which is then used as one input of an image denoising system based on SOMA [34, 36]. The second input of this system is the noisy image, I_b, and its output is the final version of the denoised image denoted as I_d^2. In an earlier work [35], this image denoising system [34, 36] required the use of the clean image, I as one input. However, this image, I, is not available, and we have only the noisy image,I_b. For that reason, we have used AFMF [8] to denoise the noisy image, I_b, to obtain a denoised image, I_d^1, which is then considered as the clean image and used as one input of this image denoising system based on SOMA [34, 36].

1.5 Results and Discussion

In this part, are presented the results obtained from the calculations of the peak signal-to-noise ratio (PSNR) and the structural similarity (SSIM). Those results were obtained by applying five image denoising approaches to a number of grey scale images. These approaches are the image denoising technique introduced in this work, the image denoising approach based on SOMA [34, 36], the universal the thresholding-based denoising approach [36], the deep learning-based denoising approach [37], and the image denoising approach based on TV-L1 model optimized with a primal–dual algorithm [38–40]. All those previously mentioned image denoising techniques were applied to a number of noisy grey-scale images, and the obtained results are in terms of PSNR and SSIM. These results are listed in Table 1.1 and the best values of PSNR and SSIM are highlighted in bold. These best values are obtained from the application of the image denoising approach proposed in [3].

Figures 1.5, 1.6, 1.7, 1.8, 1.9, 1.10, 1.11, 1.12, 1.13 and 1.14 are illustrated some examples of image denoising by employing the image denoising approach proposed in [3] and the universal thresholding-based image denoising method [36].

For ameliorating the image denoising technique introduced in [3] where the block diagram is illustrated at Fig. 1.4, we replaced the denoising system based on SOMA [34, 36] by the targeted image denoising (TID) [41, 42], which is a more recent denoising

Table 1.1 Results obtained in terms of PSNR and SSIM for the image denoising approach introduced in [3], the SOMA-based denoising approach, the universal thresholding technique, the deep learning-based denoising technique, and the image denoising approach employing TV-L1 model optimized with a primal–dual algorithm

Noisy image	The denoising approach				
	The proposed image denoising system using only the noisy image as input [3]	SOMA-based denoising technique [34, 36]	The universal thresholding-based denoising technique [36]	Deep learning-based denoising technique [37]	Image denoising technique using TV-L1 model optimized with a primal–dual algorithm [38–40]
Noisy "lena512.bmp" (*PSNRI* (*dB*) = 66.5625)	PSNR(dB): 75.7732 SSIM. 0.9738	PSNR(dB): 75.6473 SSIM: 0.9730	PSNR(dB): 68.7808 SSIM: 0.9455	PSNR(dB): 72.7664 SSIM: 0.9505	PSNR (dB): 73.2795 SSIM: 0.9531
Noisy "cameraman.tif" (*PSN*(*dB*) = 66.2173, *SSIM* = 0.8796)	PSNR (dB): 72.0261 SSIM: 0.9634	PSNR (dB): 71.9333 SSIM: 0.9624	PSNR (dB): 68.2102 SSIM: 0.9187	PSNR (dB): 71.5820 SSIM: 0.9609	PSNR (dB): 71.5820 SSIM: 0.9474
Noisy "boat.tif" (*PSN*(*dB*) = 66.7000, SSIM = 0.8211)	PSNR(dB): 73.7129 SSIM: 0.9562	PSNR (dB): 73.6766 SSIM: 0.9559	PSNR (dB): 70.8314 SSIM: 0.9194	PSNR (dB): 72.9149 SSIM: 0.9492	PSNR(dB): 71.3264 SSIM: 0.9211
Noisy "coco.tif" (*PSNRi* (*dB*) = 66.6648, SSIM = 0.8158)	PSNR(dB): 76.9281 SSIM: 0.9785	PSNR (dB): 76.8794 SSIM: 0.9779	PSNR (dB): 68.2658 SSIM: 0.8643	PSNR (dB): 73.5074 SSIM: 0.9547	PSNR (dB): 74.6719 SSIM: 0.9634
Noisy "al.tif" (*PSNRi* (*dB*) = 65.9736, SSIM = 0.9013)	PSNR (dB): 75.7546 SSIM: 0.9884	PSNR (dB): 75.3402 SSIM: 0.9872	PSNR (dB): 68.7657 SSIM: 0.9453	PSNR (dB): 71.0639 SSIM: 0.9670	PSNR(dB): 72.9949 SSIM: 0.9781
Noisy "mit.tif" (*PSNRi* (*dB*) = 65.2822, SSIM = 0.8951)	PSNR (dB): 70.4586 SSIM: 0.9640	PSNR(dB): 70.3055 SSIM: 0.9626	PSNR(dB): 67.0341 SSIM: 0.9261	PSNR (dB): 69.7257 SSIM: 0.9591	PSNR(dB): 69.3681 SSIM: 0.9519
Noisy "man.tif" (*PSNRi* (*dB*) = 66.4126, SSIM = 0.8166)	PSNR (dB): 74.3454 SSIM: 0.9631	PSNR(dB): 70.2698 SSIM: 0.9622	PSNR(dB): 69.7689 SSIM: 0.9037	PSNR(dB): 71.9095 SSIM: 0.9393	PSNR (dB): 71.8661 SSIM: 0.9340

Fig. 1.5 Noisy image (original image ("lena512.bmp [44]") corrupted by salt and pepper noise (PSNR = 66.5625, SSIM = 0.8233)

Fig. 1.6 Denoised image obtained by the proposed image denoising approach (PSNR = 75.7732, SSIM = 0.9634)

approach. The latter is an image denoising system that is also requiring the employment of the clean image, I, as one input and the corresponding noisy image, I_b, as its second input. Though, a clean image is not available, and for this reason, the AFMF-based denoising method [8] was firstly applied to the noisy image, I_b, for having the first version of the denoised image, I_d^1. The latter was considered as a clean image. After that, the TID system

Fig. 1.7 Denoised image obtained by universal thresholding [36] (PSNR = 69.6857, SSIM = 0.9041)

Fig. 1.8 Noisy image (original image ("cameraman.tif" [44]) corrupted by salt and pepper noise (PSNR = 66.2173, SSIM = 0.8796)

Fig. 1.9 Denoised image obtained by universal thresholding [36] (PSNR = 68.2102, SSIM = 0.9187)

Fig. 1.10 Denoised image obtained by the proposed image denoising approach (PSNR = 72.0261, SSIM = 0.9634)

Fig. 1.11 Noisy image (original image "boat.tif") corrupted by salt and pepper noise (PSNR = 66.7000, SSIM = 0.8211)

Fig. 1.12 Denoised image obtained by universal thresholding [36] (PSNR = 70.8314, SSIM = 0.9194)

Fig. 1.13 Denoised image obtained by the proposed image denoising approach (PSNR = 73.7129, SSIM = 0.9562)

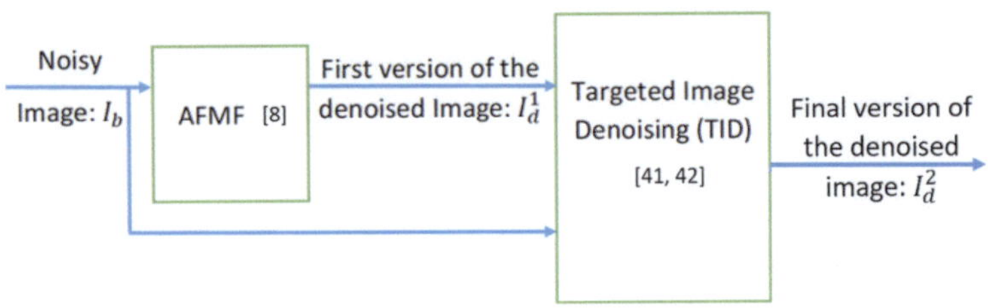

Fig. 1.14 The block diagram of the second and improved image denoising technique introduced in [3]

[41, 42] has been applied to the two inputs, I_b and I_d^1 for finally having the final version of the denoised image, I_d^2, as the output of this system [41, 42]. The block diagram of this modified image denoising approach introduced in [3] is shown in Fig. 1.14.

For studying the performance of the ameliorated image denoising technique [3] (Fig. 1.14), it was compared to the AFMF-based denoising technique [8]. This comparison was made by calculating the PSNR and SSIM and the obtained results are listed in Table 1.2 [3].

According to Table 1.2, one can remark that the ameliorated image denoising technique (Fig. 1.14) [3] outperforms the AFMF-based denoising approach [8].

Table 1.2 Results obtained in terms of PSNR and SSIM for the proposed modified image denoising technique and the AFMF-based image denoising technique

	The image denoising technique	
The noisy and the corresponding clean images	The improved image denoising technique [3]	The AFMF-based denoising technique [8]
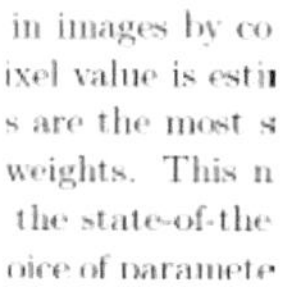 The clean image 1 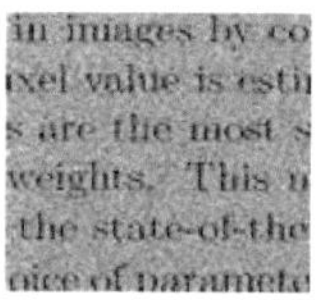 The noisy image 1: PSNR = 60.7890 SSIM = 0.6050  The clean image 2 [43] 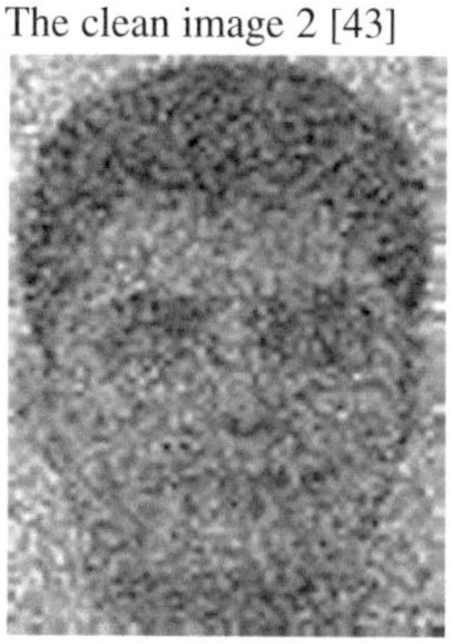 The noisy image 2: PSNR = 60.8615 SSIM = 0.6148	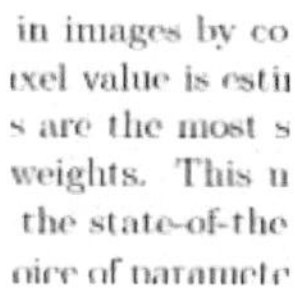 The denoised image 1: PSNR = 72.8184 SSIM = 0.9574  The denoised image 2: PSNR = 73.8984 SSIM = 0.9669	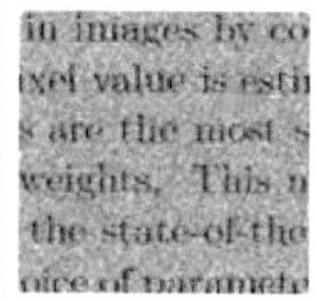 The denoised image 1: PSNR = 63.2341 SSIM = 0.6906 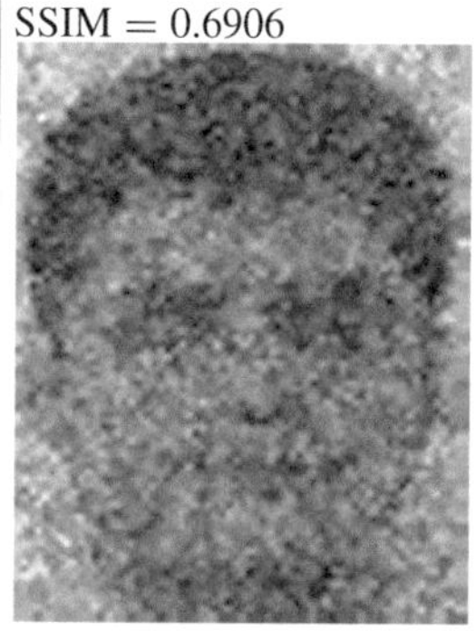 The denoised image 2: PSNR = 63.9919 SSIM = 0.7609

1.6 Conclusion

In this chapter, we detailed an image denoising technique introduced in [3]. This technique is combining the adaptive frequency median filter (AFMF) and the self-organizing migration algorithm (SOMA). This technique involves firstly applying the AFMF to the noisy image for having the first version of the denoised image. The latter is considered as a clean image and employed as an input of an image denoising system employing the SOMA. The latter has two inputs which are the clean image and the corresponding noisy image. After that, this denoising system is applied for denoising the noisy image, and a final version of the denoised image, is obtained. In reality, the clean image is not available and has to be determined. For this reason, we firstly applied the AFMF to the noisy image in order to obtain a first version of the denoised image and consider it as a clean image. The performance of this technique has been proven by the results obtained by calculating the peak signal-to-noise ratio (PSNR) and structural similarity (SSIM). These results have shown that the proposed image denoising technique outperforms a number of denoising techniques applied in this work for our evaluation, and they are as follows: the image denoising technique based on SOMA [34, 36], the universal thresholding-based denoising method [36], the deep learning-based denoising technique [37], and the image denoising method employing TV-L1 model optimized with a primal–dual algorithm [38–40]. For amliorating this proposed denoising technique, we have replaced the denoising system based on SOMA [34, 36] with the targeted image denoising (TID) [41, 42], which is a more recent denoising approach. For studying the performance of this improved image denoising technique [3], we have compared it to the AFMF-based image denoising technique [8]. This comparison has been made by calculating the PSNR and SSIM, and the results obtained from these comptations, have shown that this improved image denoising technique outperforms the AFMF-based image denoising method [8].

Abbreviations List

TID Targeted image denoising
SOMA Self-organizing migration algorithm
AFMF Adaptive frequency median filter
ML Migration loop

References

1. Thanh, D. N. H., Prasath, V. B. S., Son, N. V., & Hieu, L. M. (2019). An adaptive image inpainting method based on the modified mumford shah model and multiscale parameter estimation. *Komput Opt, 43*(2), 251–257. https://doi.org/10.18287/2412-6179-2019-43-2-251-257

2. Erkan, U., Gökrem, L., & Enginoğlu, S. (2018). Different applied median filter in salt and pepper noise. *Computers and Electrical Engineering, 70*, 789–798. https://doi.org/10.1016/j.compeleceng.2018.01.019
3. Talbi, M., Nasraoui, B. (2024). Novel image denoising techniques using AFMF. *Recent Advances in Electrical & Electronic Engineering, 17*(5). https://doi.org/10.2174/0123520965262527231218032707
4. Sethunadh, R., & Thomas, T. (2015). Spatially adaptive image denoising using inter-scale dependence in directionlet domain. *IET Image Process., 9*(2), 142–152. https://doi.org/10.1049/iet-ipr.2014.0112
5. Jennifer Ranjani, J., & Chithra, M. S. (2015). Bayesian denoising of ultra sound images using heavy-tailed levy distribution. *IET Image Processing, 9*(4), 338–345. https://doi.org/10.1049/iet-ipr.2013.0863
6. Wang, X., Chen, W., Gao, J., & Wang, C. (2018). Hybrid image denoising method based on non-subsampled contourlet transform and bande let transform. *IET Image Process., 12*(5), 778–784. https://doi.org/10.1049/iet-ipr.2017.0647
7. Guo, F., Zhang, C., & Zhang, M. (2018). Edge-preserving image de noising. *IET Image Processing, 12*(8), 1394–1401. https://doi.org/10.1049/iet-ipr.2017.0880
8. Erkan, U., Enginoğlu, S., Thanh, D. N. H., & Hieu, L. M. (2020). Adaptive frequency median filter for the salt and pepper denoising problem. *IET Image Processing, 14*(7), 1291–1302. https://doi.org/10.1049/iet-ipr.2019.0398
9. Mafi, M., Rajaei, H., Cabrerizo, M., & Adjouadi, M. (2018). A robust edge detection approach in the presence of high impulse noise in tensity through switching adaptive median and fixed weighted mean filtering. *IEEE Transactions on Image Processing, 27*(11), 5475–5490. https://doi.org/10.1109/TIP.2018.2857448. PMID:30028707.
10. Wang, X., Shen, S., Shi, G., Xu, Y., & Zhang, P. (2016). Iterative non local means filter for salt and pepper noise removal. *Journal of Visual Communication Image Represention, 38*, 440–450. https://doi.org/10.1016/j.jvcir.2016.03.024
11. González-Hidalgo, M., Massanet, S., Mir, A., & Ruiz-Aguilera, D. (2018). Improving salt and pepper noise removal using a fuzzy mathemat ical morphology-based filter. *Applied Soft Computing, 63*, 167–180. https://doi.org/10.1016/j.asoc.2017.11.030
12. Prasath, S., & Thanh, D. N. (2019). Structure tensor adaptive total varia tion for image restoration. *Turkish Journal of Electrical Engineering and Computer Sciences, 27*(2), 1147–1156. https://doi.org/10.3906/elk-1802-76
13. Dahl, J., Hansen, P. C., Jensen, S. H., & Jensen, T. L. (2010). Algorithms and software for total variation image reconstruction via first-order methods. *Numerical Algorithms, 53*(1), 67–92. https://doi.org/10.1007/s11075-009-9310-3
14. Le, T., Chartrand, R., & Asaki, T. J. (2007). A variational approach to reconstructing images corrupted by Poisson noise. *Journal of Mathematical Imaging and Vision, 27*(3), 257–263. https://doi.org/10.1007/s10851-007-0652-y
15. Thanh, D., Surya, P., & Hieu, L. M. (2019). A review on CT and x-ray images denoising methods. *Informatica (Vilnius), 43*, 151–159
16. Thanh, D. N. H., & Dvoenko, S. D. (2016). A method of total variation to remove the mixed Poisson-Gaussian noise. *Pattern Recognition Image Analysis, 26*(2), 285–293. https://doi.org/10.1134/S1054661816020231
17. Enginoğlu, S., Erkan, U., & Memiş, S. (2019). Pixel similarity-based adaptive Riesz mean filter for salt-and-pepper noise removal. *Multimedia Tools Application, 78*(24), 35401–35418. https://doi.org/10.1007/s11042-019-08110-1
18. Chen, T., & Wu, H. R. (2001). Adaptive impulse detection using center weighted median filters. *IEEE Signal Processing Letters, 8*(1), 1–3. https://doi.org/10.1109/97.889633

19. Azhar, M., Dawood, H., Dawood, H., Choudhary, G. I., Bashir, A. K., & Chauhdary, S. H. (2019). Detail-preserving switching algorithm for the removal of random-valued impulse noise. *Journal of Ambient Intelligence and Humanized Computing, 10*(10), 3925–3945. https://doi.org/10.1007/s12652-018-1153-0
20. Iqbal, N., Ali, S., Khan, I., & Lee, B. (2019). Adaptive edge preserving weighted mean filter for removing random-valued impulse noise. *Symmetry (Basel), 11*(3), 395. https://doi.org/10.3390/sym11030395
21. Chen, J., Zhan, Y., Cao, H., & Wu, X. (2018). Adaptive probability filter for removing salt and pepper noises. *IET Image Processing, 12*(6), 863–871. https://doi.org/10.1049/iet-ipr.2017.0910
22. Sheik Fareed, S. B., & Khader, S. S. (2018). Fast adaptive and selective mean filter for the removal of high-density salt and pepper noise. *IET Image Processing, 12*(8), 1378–1387. https://doi.org/10.1049/iet-ipr.2017.0199
23. Hong, N. M., & Thanh, N. C. (2020). Distance-based mean filter for image denoising. In *Proceeding 4th International Conference on Machine Learning and Soft Computing*, HaiPhong City, Vietnam. https://doi.org/10.1145/3380688.3380704
24. Erkan, D. N. H. T, Hieu, L. M, & Enginoglu, S. (2019). An iterative mean filter for image denoising. *IEEE Access, 7*(1), 167847–167859. https://doi.org/10.1109/ACCESS.2019.2953924
25. Thanh, D. N. H., Thanh, L. T., Hien, N. N., & Prasath, S. (2020). Adaptive total variation L1 regularization for salt and pepper image de noising. *Optik (Stuttg.), 208*, 163677. https://doi.org/10.1016/j.ijleo.2019.163677
26. Tukey, J. W. (1977). Exploratory data analysis, vol. 2. Addison-Wesley Publishing Company: Boston
27. Varatharajan, R., Vasanth, K., Gunasekaran, M., Priyan, M., & Gao, X. Z. (2018). An adaptive decision based kriging interpolation algorithm for the removal of high density salt and pepper noise in images. *Computers and Electrical Engineering, 70*, 447–461. https://doi.org/10.1016/j.compeleceng.2017.05.035
28. Fu, B., Zhao, X., Song, C., Li, X., & Wang, X. (2019). A salt and pepper noise image denoising method based on the generative classifica tion. *Multimedia Tools Application, 78*(9), 12043–12053. https://doi.org/10.1007/s11042-018-6732-8
29. Ko, S. J., & Lee, Y. H. (1991). Center weighted median filters and their applications to image enhancement. *IEEE Transaction Circuits System, 38*(9), 984–993. https://doi.org/10.1109/31.83870
30. Brownrigg, D. R. K. (1984). The weighted median filter. *Communications of the ACM, 27*(8), 807–818. https://doi.org/10.1145/358198.358222
31. Yin, L., Yang, R., Gabbouj, M., & Neuvo, Y. (1996). Weighted median filters: A tutorial. *IEEE Transitions on Circuits System, 2 Analog Digital Signal Processing, 43*(3), 157–192. https://doi.org/10.1109/82.486465
32. Thanh, D. N. H., Thanh, L. T., & Prasath, S. (2019). An improved BPDF filter for high density salt and pepper denoising. In *Proceeding of IEEE RIVF International Conference on Computing and Communication Technologies*, Danang. https://doi.org/10.1109/RIVF.2019.8713669
33. Sree, P. S. J., Kumar, P., Siddavatam, R., & Verma, R. (2013). Salt-and pepper noise removal by adaptive median-based lifting filter using second-generation wavelets. *Signal Image Video Processing, 7*(1), 111–118. https://doi.org/10.1007/s11760-011-0210-3
34. Cao, Z., Jia, H., Zhao, T., Fu, Y., & Wang, Z. (2022). An adaptive self organizing migration algorithm for parameter optimization of wavelet transformation. *Mathematical Problems Engineering, 2022*. https://doi.org/10.1155/2022/6289215
35. Zelinka, I. (2004). Soma–self-organizing migrating algorithm. *Studies in Fuzziness and Soft Computing, 141*, 167–217. https://doi.org/10.1007/978-3-540-39930-8_7

36. MathWorks, Image denoising using Evolutionary Algorithm–File Exchange–MATLAB Central, Available From: https://www.mathworks.com/matlabcentral/fileexchange/46649-image-denoising-using-evolutionary-algorithm/
37. Zhang, K., Zuo, W., Chen, Y., Meng, D., & Zhang, L. (2017). Beyond a Gaussian Denoiser: residual learning of deep CNN for image denoising. *IEEE Transitions Image Processing, 26*(7), 3142–3155. https://doi.org/10.1109/TIP.2017.2662206 PMID: 28166495
38. Lourakis, M. TV-L1 image denoising algorithm, Available From: https://www.mathworks.com/matlabcentral/fileexchange/57604tv-l1-image-denoising-algorithm
39. Mordvintsev, A. ROF and TV-L1 denoising with primal-dual algorithm, Available From: http://znah.net/rof-and-tv-l1denoising-with-primal-dual-algorithm.html
40. Chambolle, A. (2009). An introduction to total variation for image analysis. *Radon Series Comp. Appl. Math., 9*, 263–340.
41. Luo, E., Chan, S. H., & Nguyen, T. Q. (2015). Adaptive image denoising by targeted databases. *IEEE Transactions on Image Processing, 24*(7), 2167–2181. https://doi.org/10.1109/TIP.2015.2414873. PMID: 25807566.
42. Luo, E., Chan, S. H., & Nguyen, T. Q. Image denoising by targeted external databases. In *2014 IEEE International Conference on Acoustics, Speech and Signal Processing (ICASSP)*, 2014 Florence, Italy. https://doi.org/10.1109/ICASSP.2014.6854040
43. MathWorks, TID–Targeted image denoising–File exchange–MATLAB central, Available From: https://www.mathworks.com/matlabcentral/fileexchange/55776tid-targeted-image-denoising?tab=discussions
44. Zhang, L., Dong, W., Shi, G., & Zhang, D. (2010). Two-stage image denoising by principal component analysis with local pixel grouping. *Pattern Recognition, 43*(4), 1531–1549.

An Image Denoising Technique Based on 2D Dual-Tree DWT and Stationary Wavelet Transform 2D

2

2.1 Introduction

Noisy images often arise in high level vision tasks and these make image denoising an important task in the low level vision domain [1]. For example, take a given denoising model:

$$y = x + n. \tag{2.1}$$

With y, n, and x are respectively the noisy image, the additive white Gaussian noise (AWGN) and the corresponding clean image. This noise is with σ as the standard deviation [1]. There are different denoising techniques proposed in the literature and each of them owns advantages and drawbacks. In [2], a review of some interesting research works in the image denoising domain was performed, and based on that, the denoising methods were presented. The methods can be classified as wavelet, spatial domains, or both methods can be combined for having the advantages. In the medical domain, Mittal et al. [3] have proposed a methodology for ameliorating and cancelling the high noise of the medical image via the employment of the stationary wavelet transform (SWT) method. In their work, an effective and simple technique for adaptive noise cancellation was used, the SWT-2D-based denoising approach on the noisy medical image. In [4], our image denoising technique proposed in literature consists in applying a SWT-based image denoising method [5] in the domain of 2D dual-tree discrete wavelet transform (DWT) [6]. This SWT-based image denoising method is based on denoising the noisy wavelet coefficients obtained from the noisy image decomposition by employing the SWT-2D. This denoising is performed by employing the soft thresholding. For this noisy image decomposition, the Daubechies 4 is used as the mother wavelet and 5 is the level of this decomposition. These choices are the same as those in [5]. The rest of this chapter is arranged as follows:

© The Author(s), under exclusive license to Springer Nature Switzerland AG 2026

M. Talbi and B. Nasraoui, *Image Denoising Using Mathematical Tools*, Synthesis Lectures on Engineering, Science, and Technology,
https://doi.org/10.1007/978-3-032-11878-3_2

Part 2 devoted to 2D dual-tree complex wavelet transform [6]. In part 3, we will deal with the SWT-2D-based image denoising method [5]. In part 4, we will detail our image denoising approach introduced in [4]. In part 5, we will present results and discussion, and finally, the conclusion will be given in part 6.

2.2 2D Dual-Tree Complex Wavelet Transform

For some DWT applications, improvements can be obtained by using an expansive wavelet transform in place of a critically sampled one [6]. An expansive transform is one that allows to convert an N-point signal into M coefficients where $M > N$. Many types of expansive DWTs, and among them we have the dual-tree complex DWT [6, 7]. As shown in Fig. 2.1, the dual-tree complex DWT of a signal x is performed by using two critically sampled DWTs in parallel on the same data.

This transform is two times expansive due to the fact that for an N-point signal, it permits to obtain $2N$DWT coefficients. In case where the filters in both lower and upper DWTs are the same, therefore no supplimentary advantage. Though, in case where the filters are designed in a specific manner, therefore, the sub-band signals of the upper DWT can be viewed as the real part of a complex wavelet transform, and sub-band signals of the lower DWT can be considered as the imaginary part. Equivalently, for specially designed filter sets, the wavelet associated with the upper DWT can be considered as an approximate Hilbert transform of the wavelet associated with the lower DWT. In case where designed in this way, the dual-tree complex DWT is approximately shift invariant, in contrary to the critically sampled DWT. Moreover, the dual-tree complex DWT can be

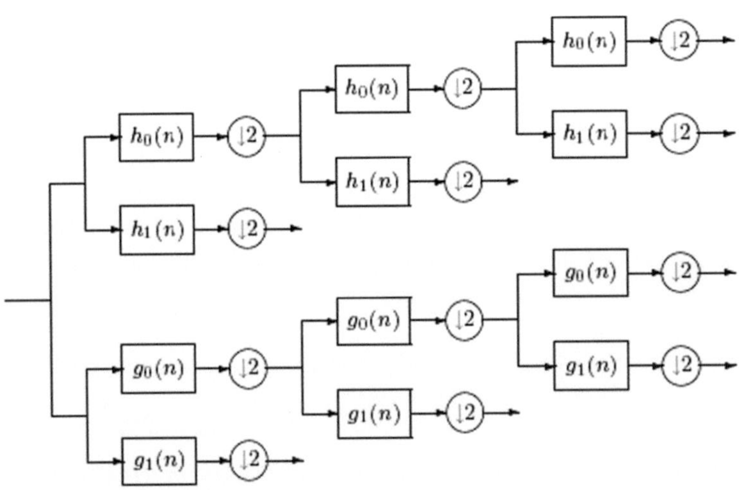

Fig. 2.1 Dual-tree complex DWT

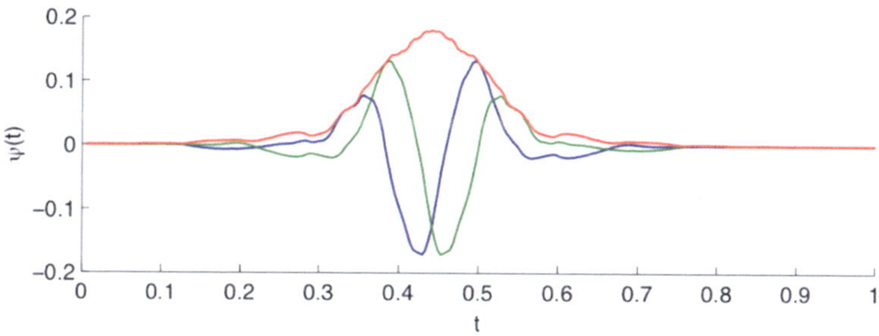

Fig. 2.2 Complex 1D wavelet, $\psi(t)$ [6, 7]

used for the implementation of the 2D wavelet transforms where each wavelet is oriented, which is particularly useful for image processing. For the separable 2D DWT, recall that one of the three wavelets hasn't a dominant orientation. The dual-tree complex DWT is outperforming the critically sampled DWT for applications such as image denoising. The complex wavelet associated with the dual-tree complex is shown in Fig. 2.2.

For determining the real part of the complex wavelet, all coefficients are setted to 0, except for one coefficient in the upper DWT, then the inverse transform is applied. In order to have the imaginary part, the exception is a single coefficient in the lower DWT.

2.3 A SWT-Based Image Denoising Method

The SWT [8, 9] is similar to the discrete wavelet transform (DWT) except that the signal is never sampled and instead the filters are up sampled at each of the decomposition levels [8]. Each level's filters are up sampled versions of the previous, as illustrated in Fig. 2.3.

The SWT is an inherent redundant scheme, as each coefficients set is containing the same number of samples as the input. Consequently, for a decomposition of N levels, there is a $2N$ redundancy. An image denoising method by applying the soft thresholding in the SWT field can be summarized by the block diagram shown in Fig. 3.4. This denoising method is applied in our denoising system previously proposed in literature [4].

Following Fig. 2.4, the SWT-2D is firstly applied to the noisy image, I_b, for having noisy stationary wavelet coefficients. These obtained coefficients are then denoised by using soft thresholding, and finally, the inverse of SWT – 2D which is SWT^{-1} – 2D is applied to the obtained thresholded coefficients in order to have the denoised image, I_d.

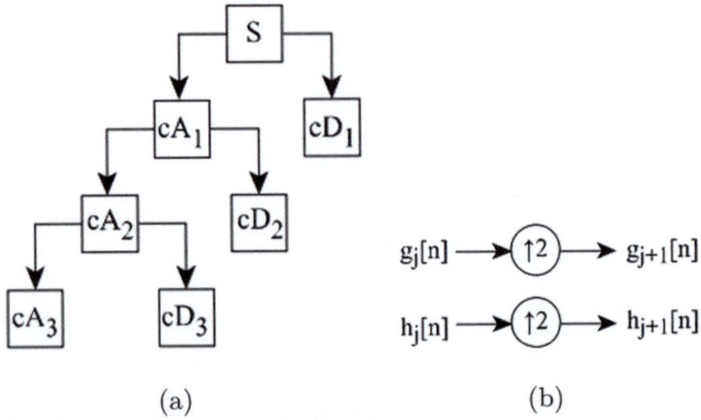

Fig. 2.3 **a** Wavelet decomposition tree, **b** SWT filters [8]

Fig. 2.4 Block diagram of an image denoising method by applying the soft thresholding in the SWT domain

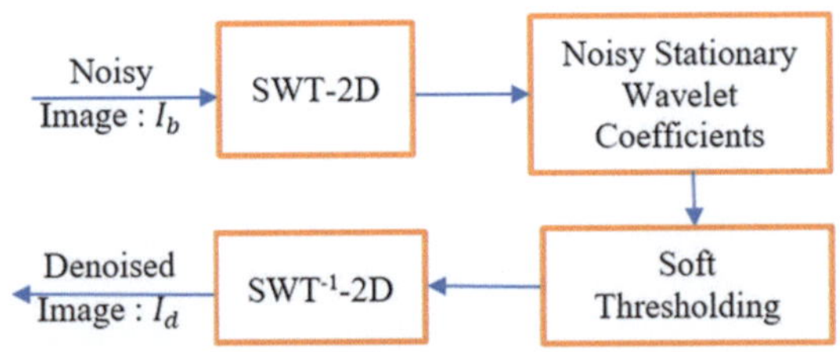

2.4 The Image Denoising Approach Proposed in [4]

In this section, we will detail the image denoising approach proposed in [4]. It is based on applying a SWT-based image denoising method in the 2D dual-tree DWT domain [4]. In fact, our denoising approach proposed in [4] consists firstly in applying the 2D dual-tree DWT to the noisy image, I_b in order to have noisy wavelet coefficients. The latters are secondly denoised by applying a SWT-based image denoising method [5]. Finally, the denoised image is obtained from the application of the inverse of 2D dual-tree DWT to the denoised wavelet coefficients obtained in the second step. This image denoising technique introduced in [4] can be summarized by the block diagram shown in Fig. 2.5.

According to Fig. 2.5, the 2D dual-tree DWT is firstly applied to the noisy image, I_b, for having noisy wavelet coefficients, denoted as Wb{j}{s}, $1 \leq j \leq 2, 1 \leq s \leq 3$. After that, each of these coefficients is denoised through the application of an image denoising technique based on thresholding in SWT domain [5], for having denoised wavelet coefficients, denoted as Wd{j}{s}, $1 \leq j \leq 2, 1 \leq s \leq 3$. Finally, the inverse of the 2D dual-tree DWT is applied to these denoised coefficients for having the denoised image, denoted as I_d. Consequently, in [4], was combined two efficient denoising approaches where the first is based on SWT and the second is based on dual-tree DWT (to take the advantage of

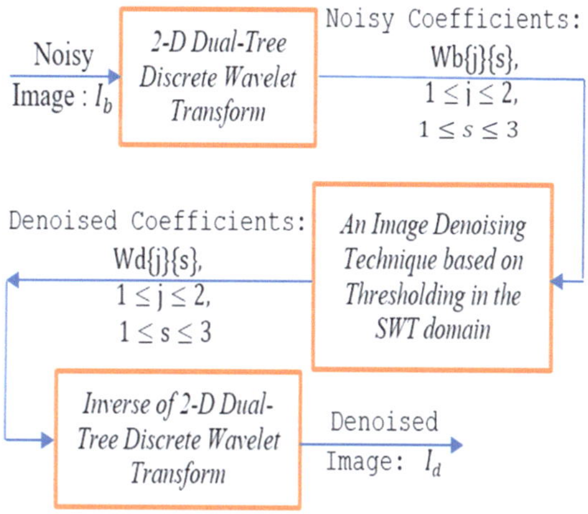

Fig. 2.5 Block diagram of our image denoising approach introduced in [4]

both). Moreover, applying the SWT denoising in the dual-tree DWT domain is providing more adaptability in comparison with just employing dual-tree DWT thresholding [4]. As previously mentioned, for applying the denoising method-based thresholding in the SWT domain, Daubechies 4 (db4) is used as the mother wavelet. Moreover, the value of the decomposition level is chosen to be equal to five. These choices are the same as those in [5]. Concerning the employment, the db4 in the image denoising approach introduced in [4]; in the wavelet transform, the mother wavelet functions have to be selected in a comprehensive manner in terms of symmetry, support length, similarity, and regularity. Those functions such as sym, db and coif, are mostly used. The db mother wavelets were introduced by Daubechies and are generally abbreviated as dbN (*N* designates the order of the mother wavelet). For a fixed number *N*, the dbN owns a good regularity and is characterized by nonlinear phase energy concentrated near the beginning of its support. Concerning the dbN wavelet, as the order (sequence N) is increasing, the greater the order of the vanishing moment, where the higher the vanishing moment, the better the smoothness, the stronger the localization capability of the frequency domain, and the better the division of the frequency band, though it will weaken the time domain tight support, while the computation volume is veneably increasing and the real-time performance becomes worse [10].

2.5 Weakness of the Image Denoising Approach Introduced in [4]

In spite of the conceptual simplicity and the straightforward implementation of our image denoising approach introduced in [4], it owns higher computational complexity in comparison with the SWT or dual-tree DWT denoising alone, due to the fact that in this approach

[4] is combining these two transforms. This combination is performed by applying SWT to each dual-tree DWT coefficient which increases computations.

2.6 Results and Discussions

In this section, the results are listed in Table 2.1 and they are obtained by calculating the peak signal-to-noise ratio (PSNR) and the structural similarity (SSIM). These results are obtained from the application of our image denoising approach introduced in [4], the denoising method based on thresholding in the SWT domain [5], the image denoising method based on thresholding in the domain of 2D dual-tree DWT [6], and the image denoising method using deep neural network [11].

Concerning the PSNR, it is a better testing due to the fact that it takes the signal strength into consideration (not only the error). The PSNR is formulated as follows [12]:

$$\mathrm{PSNR} = 10 \cdot \log_{10}\left(\frac{\mathrm{MAX}_I^2}{\mathrm{MSE}}\right). \tag{2.2}$$

With MSE is the mean square error and is formulated as follows:

$$\mathrm{MSE} = \frac{1}{n}\sum_{i=1}^{n}\left(X_i - X_i^*\right)^2. \tag{2.3}$$

Concerning the SSIM, its value is in general in the range, [0, 1]. Similar images own a SSIM value near to 1. The SSIM is formulated as follows:

$$\mathrm{SSIM} = \frac{(2\mu_x\mu_y + c_1)(2\sigma_{xy} + c_2)}{\left(\mu_x^2 + \mu_y^2 + c_1\right)\left(\sigma_x^2 + \sigma_y^2 + c_2\right)}. \tag{2.4}$$

With σ, σ_{xy} and μ designate respectively the variance, the covariance of the image, and c_1 and c_2 are the stabilizing constants.

Also those results were obtained for diverse values of σ, the level of noise corrupting the original image (the clean image). This noise is an additive Gaussian white noise (AGWN). According to Table 2.1, the best results were highlighted in bold and they were almost obtained by applying our image denoising technique introduced in [4]. Consequently, the introduced image denoising approach [4] achieves the best denoising performance in those evaluations. In fact, this approach permits to have higher values of PSNR and SSIM and this in comparison to other image denoising techniques such as the denoising technique based on soft thresholding in SWT 2D domain [5], the denoising technique based on soft thresholding in the dual-tree DWT domain [6], and the denoising method based on deep learning [11].

Table 2.1 Result obtained in terms of PSNR and SSIM

Noisy image	The denosiong technique			
	The proposed image denoising technique [4]	The image denoising technique based on thresholding in the SWT domain [5]	The image denoising technique based on deep neural network [11]	The image denoising technique based on soft thresholding in the domain of 2D dual-tree DWT [6]
Noisy image, 'st.tif' (σ = 10)	PSNR: 34.2517 SSIM: 0.8791	PSNR: 33.7831 SSIM: 0.8962	PSNR: 28.3030 SSIM: 0.5425	PSNR: 34.1453 SSIM: 0.8675
Noisy st.tif (σ = 20)	PSNR: 30.6126 SSIM: 0.7722	PSNR: 30.1884 SSIM: 0.6950	PSNR: 22.4725 SSIM: 0.2850	PSNR: 30.2840 SSIM: 0.7306
Noisy st.tif (σ = 30)	PSNR: 28.1244 SSIM: 0.6131	PSNR: 24.5706 SSIM: 0.3719	PSNR: 18.9533 SSIM: 0.1789	PSNR: 27.7900 SSIM: 0.5977
Noisy st.tif (σ = 40)	PSNR: 26.4861 SSIM: 0.5138	PSNR: 20.6093 SSIM: 0.2171	PSNR: 16.4559 SSIM: 0.1248	PSNR: 25.9906 SSIM: 0.4953
Noisy Peppers.tif (σ = 10)	PSNR: 32.8048 SSIM: 0.8250	PSNR: 32.4622 SSIM: 0.8019	PSNR: 28.6273 SSIM: 0.6497	PSNR: 32.6519 SSIM: 0.8180
Noisy Peppers.tif (σ = 20)	PSNR: 29.6501 SSIM: 0.7233	PSNR: 29.3674 SSIM: 0.6821	PSNR: 22.4648 SSIM: 0.3569	PSNR: 29.5645 SSIM: 0.7038
Noisy Peppers.tif (σ = 30)	PSNR: 27.8670 SSIM: 0.6406	PSNR: 27.8481 SSIM: 0.6264	PSNR: 18.8973 SSIM: 0.2208	PSNR: 27.4151 SSIM: 0.5968
Noisy Peppers.tif (σ = 40)	PSNR: 26.3256 SSIM: 0.5462	PSNR: 26.2747 SSIM: 0.5424	PSNR: 16.3708 SSIM: 0.1497	PSNR: 25.7770 SSIM: 0.5109
Noisy House.tif (σ = 10)	PSNR: 35.0397 SSIM: 0.9203	PSNR: 34.2284 SSIM: 0.8466	PSNR: 28.4578 SSIM: 0.5576	PSNR: 35.0360 SSIM: 0.9121
Noisy House.tif (σ = 20)	PSNR: 30.7882 SSIM: 0.8462	PSNR: 30.3082 SSIM: 0.7360	PSNR: 22.4921 SSIM: 0.3090	PSNR: 30.5116 SSIM: 0.7766
Noisy House.tif (σ = 30)	PSNR: 28.1921 SSIM: 0.6586	PSNR: 27.0232 SSIM: 0.5584	PSNR: 18.9670 SSIM: 0.2049	PSNR: 27.9388 SSIM: 0.6487
Noisy House.tif (σ = 40)	PSNR: 26.2186 SSIM: 0.5368	PSNR: 25.4000 SSIM: 0.4792	PSNR: 16.4148 SSIM: 0.1455	PSNR: 26.0214 SSIM: 0.5416

Figure 2.6a–f show some examples of image denoising by applying the image denoising approach introduced in [4] and the other two techniques [6, 11] applied for the performance evaluation in [4].

According to Fig. 2.6a–f, visually, the denoised image obtained by applying the introduced denoising approach retaining more details and looks less noisy in comparison with the images denoised by the other three denoising methods [5, 6, 11]. Hence, our denoising approach previously introduced in [4] permits to have good noise reduction while conserving the original image structure. Concerning the decomposition level, it can be varied

Fig. 2.6 Example of image denoising: **a** Clean image, **b** noisy image ($\sigma = 20$), **c** denoised image obtained by applying our denoising approach introduced in [4] (PSNR = 30.6924), **d** denoised image obtained by applying deep learning [11] (PSNR = 22.4978), **e** denoised image obtained by applying soft thresholding in the SWT 2D domain [5] (PSNR = 30.1884), **f** denoised image obtained by applying the soft thresholding in the domain of the 2D dual-tree DWT [6] (PSNR = 30.2840)

Table 2.2 Results in terms of PSNR and SSIM: Noisy image ("einstein.tif" degraded by an AGWN with noise level, $\sigma = 10$)

Mother wavelet	Level 1	Level 2	Level 3	Level 4	Level 5
db4	PSNR: 32.4466 SSIM: 0.7777	PSNR: 33.7693 SSIM: 0.8320	PSNR: 34.1574 SSIM: 0.8471	PSNR: 34.2084 SSIM: 0.8494	PSNR: **34.2500** SSIM: **0.8507**

from 1 to the maximum level. The latter is depending on the size of the signal or image to be analysed and also on the type of the employed mother wavelet. In [4], for studying the impact of the decomposition level on the denoising results, the choice of the mother wavelet was fixed at db4 and the decomposition level was varying from 1 to 5 and the obtained results were in terms of PSNR and SSIM and they are listed in Tables 2.2, 2.3, 2.4 and 2.5 as follows [4]:

According to the results listed in Tables 2.2, 2.3, 2.4 and 2.5, the best values of both PSNR and SSIM are highlighted in bold and they are obtained in case where the decomposition level is equal to 5.

Table 2.3 Results in terms of PSNR and SSIM: Noisy Image ("coco.tif" degraded by an AGWN with noise level, $\sigma = 20$)

Mother wavelet	Level 1	Level 2	Level 3	Level 4	Level 5
db4	PSNR: 27.7363 SSIM: 0.5488	PSNR: 29.9246 SSIM: 0.7152	PSNR: 30.7432 SSIM: 0.7858	PSNR: 30.9355 SSIM: 0.8027	PSNR: **30.9586** SSIM: **0.8062**

Table 2.4 Results in terms of PSNR and SSIM: Noisy image ("st.tif" degraded by an AGWN with noise level, $\sigma = 30$)

Mother wavelet	Level 1	Level 2	Level 3	Level 4	Level 5
db4	PSNR: 24.5869 SSIM: 0.3673	PSNR: 26.9989 SSIM: 0.5418	PSNR: 27.9043 SSIM: 0.6329	PSNR: 28.1022 SSIM: 0.6596	PSNR: **28.1535** SSIM: **0.6650**

Table 2.5 Results in terms of PSNR and SSIM: Noisy image ("house.tif" degraded by an AGWN with noise level, $\sigma = 40$)

Mother wavelet	Level 1	Level 2	Level 3	Level 4	Level 5
db4	PSNR: 22.0383 SSIM: 0.3293	PSNR: 24.5493 SSIM: 0.4552	PSNR: 25.3901 SSIM: 0.5208	PSNR: 25.5400 SSIM: 0.5359	PSNR: **25.6487** SSIM: **0.5423**

In the following part, are considered visual comparisons of the denoised images and this for three values of the noise level ($\sigma = 10, \sigma = 20, \sigma = 30$, and $\sigma = 40$). In fact, Figs. 2.7, 2.8 and 2.9 show three denoising examples by applying the four previously mentioned denoising approaches including our technique introduced in [4] (Fig. 2.5).

According to Fig. 2.7c and d, the image denoising approach based on thresholding in the SWT 2D domain [4] considerably permitting to eliminate the noise but some important details of the original image (Fig. 2.7a) are degraded. Although, these details are preserved in case where our image denoising approach introduced in [4] is applied. The latter is permitting to cancel an important amount of noise in comparison with the noisy image (Fig. 2.7b), but less than that eliminated by the denoising technique based on the thresholding in the SWT 2D domain [4].

According to Fig. 2.8a, c, d, and e, our image denoising approach introduced in [4], the soft thresholding in the SWT 2D domain [5], and the image denoising approach based on the soft thresholding in the domain of 2D dual-tree DWT allow the noise cancellation and preserving all the original image details (Fig. 2.8a). Though, the image denoising approach based on deep neural network [11] allows obtaining a denoised image with worse visual quality in comparison with the denoised images obtained from the application of our image denoising approach introduced in [4] and the other two previously mentioned image denoising methods [5, 6].

According to Fig. 2.9c and d, the image denoising method based on the soft thresholding in the SWT 2D domain [5] and our image denoising approach introduced in [4] allows to considerably cancel the noise. Though, the denoised image obtained by applying the proposed approach [4] owns better visual quality in comparison with the denoised image obtained by applying the image denoising method based on the soft thresholding in the SWT 2D domain [5].

The following example of image denoising by applying the image denoising technique introduced in [4], where the noisy image is obtained by adding the salt and pepper noise to the clean image which is "al.tif". This example is shown in Fig. 2.10.

The following second example of image denoising by applying our image denoising technique introduced in [4]. The noisy image is obtained by adding the Poisson noise to the clean image ("boat.tif"). This example is shown at Fig. 2.11.

Below, a third example of image denoising by applying the image denoising approach is proposed in this work. The noisy image is obtained by adding the speckle noise to the clean image which is ("mandrill.tif"). This example is illustrated in Fig. 2.12.

These examples (Figs. 2.10, 2.11 and 2.12) show the performance of the image denoising technique introduced in [4]. In fact, for those three sorts of noises (Poisson, salt and peppers and speckle), this technique allows considerably improving the PSNR and also the SSIM is slightly ameliorated.

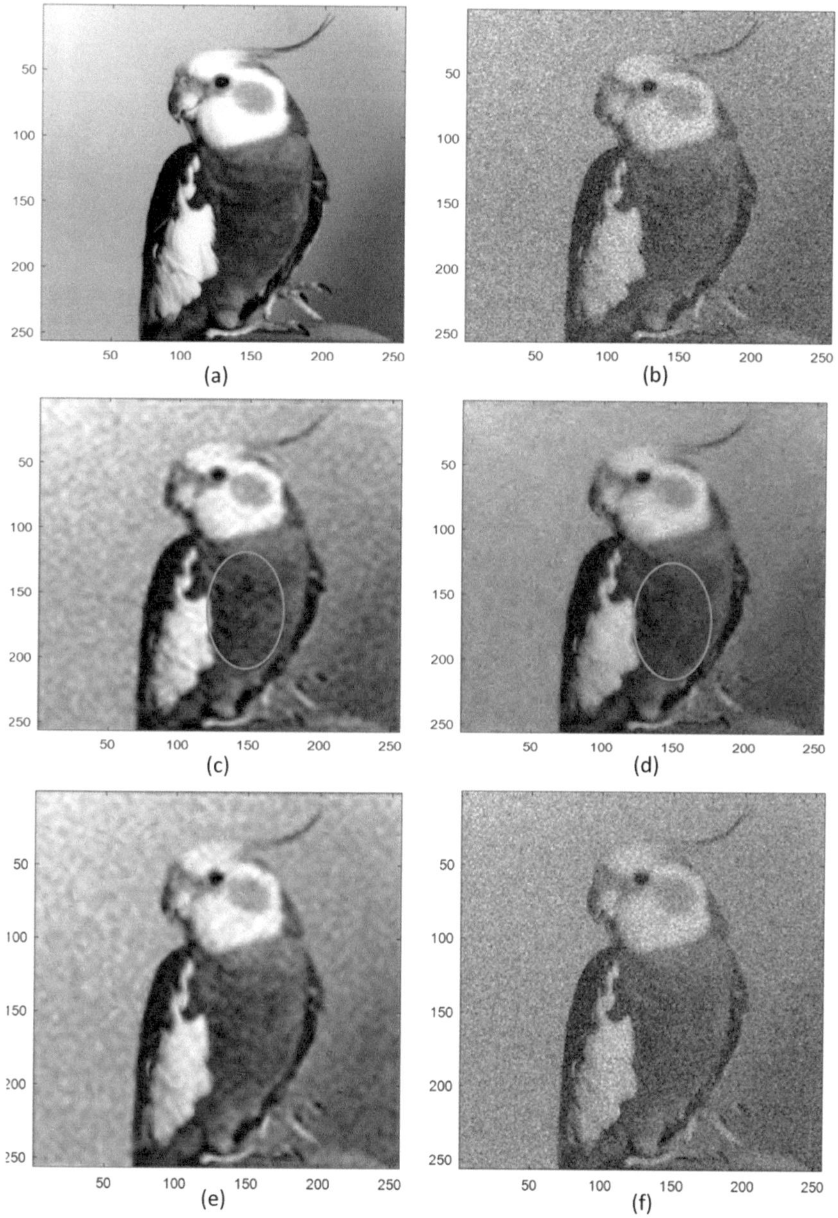

Fig. 2.7 Example of image denoising: **a** Clean image ("coco.tif"), **b** noisy image ("coco.tif" degraded by AGWN with noise level, $\sigma = 30$), **c** denoised image obtained by applying the proposed technique (PSNR = 28.7715, SSIM = 0.7100), **d** denoised image obtained by applying the soft thresholding in the SWT 2D domain (PSNR = 27.7534, SSIM = 0.5886), **e** denoised image obtained by applying the soft thresholding in the domain of 2D dual-tree DWT (PSNR = 28.5452, SSIM = 0.6854), **f** denoised image obtained by applying the deep learning (PSNR = 18.9513, SSIM = 0.1863)

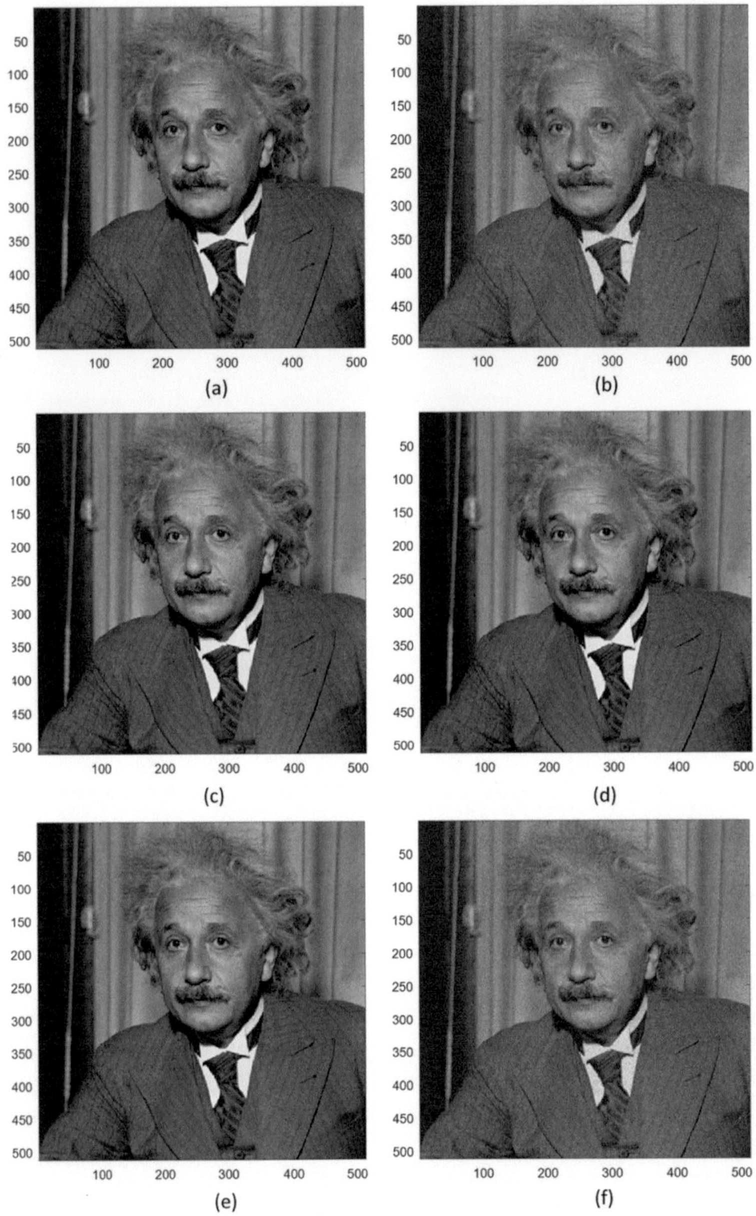

Fig. 2.8 Example of image denoising: **a** Clean image ("einstein.tif"), **b** noisy image ("einstein.tif" corrupted by an additive Gaussian white noise with noise level, $\sigma = 10$), **c** denoised image obtained by applying the proposed technique (PSNR = 34.2460, SSIM = 0.8539), **d** denoised image obtained by applying soft thresholding in the SWT domain (PSNR = 33.5834, SSIM = 0.8261), **e** denoised image obtained by applying soft thresholding in the domain of 2D dual-tree DWT (PSNR = 32.3883, SSIM = 0.8260), **f** denoised image obtained by applying deep learning (PSNR = 28.5790, SSIM = 0.6124)

Fig. 2.9 An example of image denoising: **a** Clean image ("lena.tif"), **b** noisy image (clean image ("lena.tif" corrupted by an additive Gaussian white noise with noise level, $\sigma = 40$), **c** denoised image obtained by applying the proposed technique (PSNR = 24.7147 dB, SSIM = 0.6359), **d** denoised image obtained by applying soft thresholding in the SWT domain (PSNR = 24.1291, SSIM = 0.5574), **e** denoised image obtained by applying soft thresholding in the domain of 2D dual-tree DWT (PSNR = 24.1106, SSIM = 0.5633), **f** denoised image obtained by applying deep learning (PSNR = 16.3965, SSIM = 0.2478)

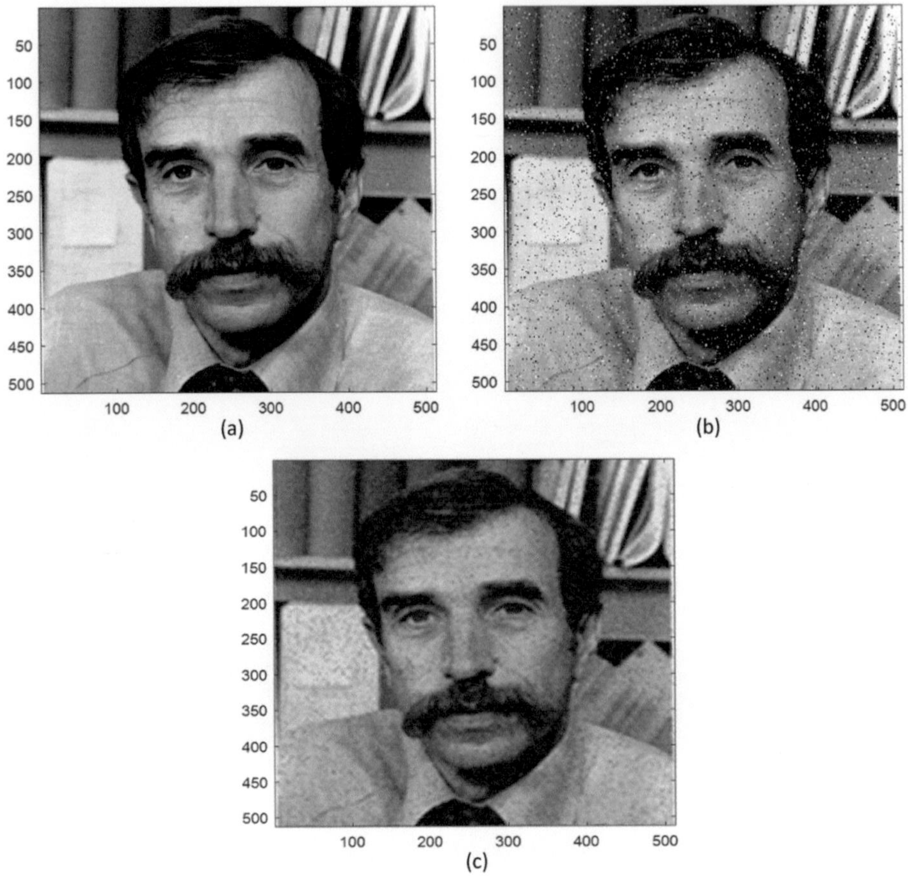

Fig. 2.10 Example of image denoising: **a** Clean image ("al.tif"), **b** noisy image (clean image ("al.tif") corrupted by an additive salt and peppers with PSNR = 65.9512 dB, SSIM = 0.9996), **c** denoised image obtained by applying the proposed technique (PSNR = 75.4904dB,SSIM = 0.9999)

2.7 Real-Time Implementation of the Proposed Image Denoising Approach

Due to the fact that our image denoising technique introduced in [4] is based on two completely different wavelet transforms which are SWT 2D and 2D dual-tree DWT, we need their real-time implementations for performing the implementation of this introduced technique. This implementation can be considered as a perspective in our future works. The real-time implementations of both SWT 2D and 2D dual-tree DWT could be performed employing the two References [13, 14].

Fig. 2.11 Example of image denoising: **a** Clean image ("boat.tif"), **b** noisy image (clean image ("boat.tif") corrupted by adding Poisson noise with PSNR = 51.0684dB, SSIM = 0.9893), **c** denoised image obtained by applying the proposed technique (PSNR = 63.6801 dB, SSIM = 0.9980)

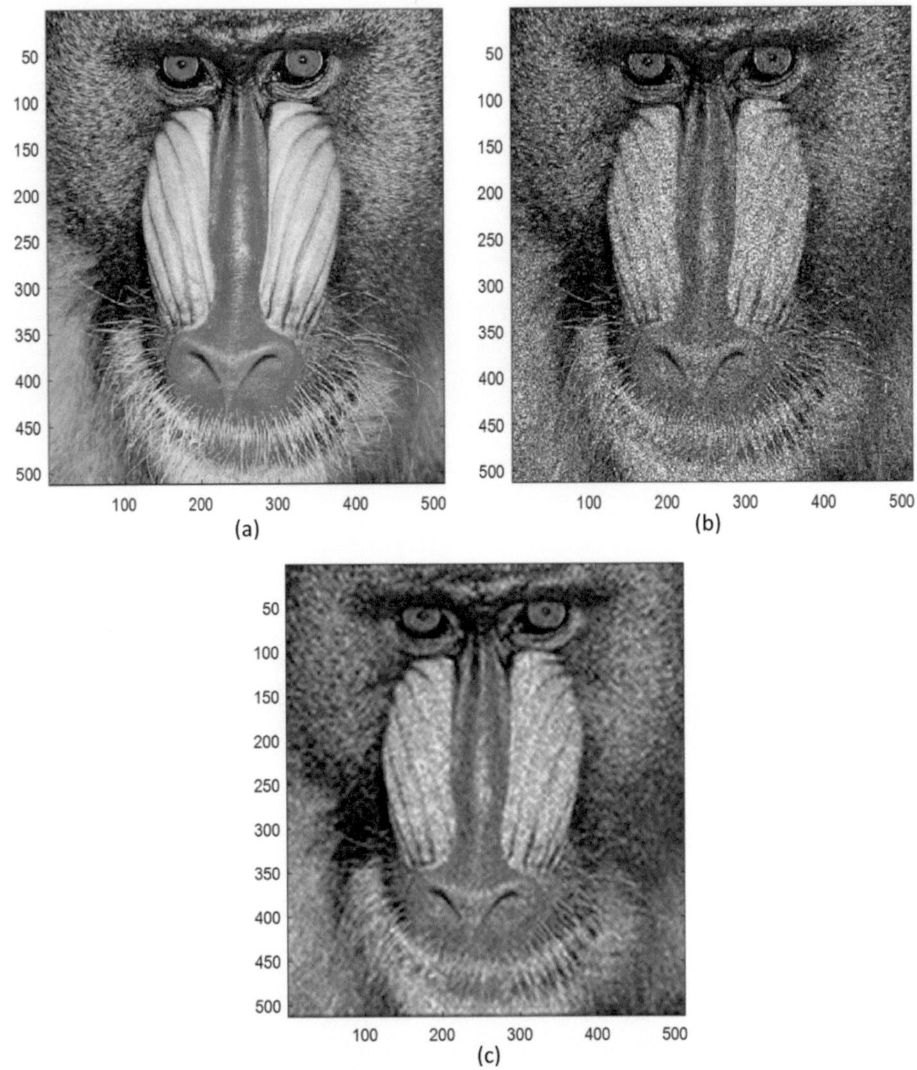

Fig. 2.12 Example of image denoising: **a** Clean image ("mandrill.tif"), **b** noisy image (clean image ("mandrill.tif") corrupted by adding Poisson noise with PSNR = 63.9702 dB, SSIM = 0.9994), **c** denoised image obtained by applying the proposed technique (PSNR = 69.1995dB, SSIM = 0.9998)

2.8 Conclusion

In this chapter, we detailed our image denoising approach introduced in literature. This technique consists in applying a SWT 2D-based image denoising method in the domain of 2D dual-tree DWT. In fact, the first step of this introduced approach consists of applying the 2D dual-tree DWT to the noisy image to obtain noisy wavelet coefficients. In the second step of this approach, these coefficients are denoised by applying to each of them a SWT 2D-based image denoising technique. Finally, the denoised image is obtained by applying the inverse of the 2D dual-tree DWT to the denoised wavelet coefficients obtained in the second step. The performance of this introduced image denoising approach is proved by the results obtained from the computation of both PSNR and SSIM. In fact, this approach allows to have the best results in terms of PSNR and SSIM and this in comparison with three other image denoising techniques existing in literature. These three techniques are the image denoising method based on neural network, the denoising method based on the soft thresholding in the SWT 2D domain and the denoising method based on thresholding in the 2D dual-tree DWT domain.

References

1. Tian, C., Fei, L., Zheng, W., Xu, Y., Zuo, W., & Lin, C. W. (2020). Deep learning on image denoising: An overview. *Neural Networks, 131*, 251–275.
2. Ismael, A. A., & Baykara, M. (2021). Digital image denoising techniques based on multi resolution wavelet domain with spatial filters: A review. *Traitement du Signal, 38*(3), 639–651. http://iieta.org/journals/ts
3. Mittal, N., Raj, A., & Goyal, H. (2019). Enhancement and removal of noise from medical images by wavelet transform method. In *3rd International conference on electronics, communication and aerospace technology*, pp. 1126–1130. https://doi.org/10.1109/ICECA.2019.8821979
4. Talbi, M., Nasraoui, B. (2024). A novel method of image denoising based on 2D dual-tree DWT and SWT. *International Journal of Wavelets, Multiresolution and Information Processing, 22*(04), 2450009
5. Generate MATLAB Code for 2D stationary wavelet denoising—MATLAB and Simulink—MathWorks France, https://www.mathworks.com/help/wavelet/ug/generating-matlab-code-for-2-d-stationary-wavelet-denoising.html
6. Wavelet software at Brooklyn poly (nyu.edu) https://eeweb.engineering.nyu.edu/iselesni/WaveletSoftware/
7. Patrick, L. (2006). Ondelettes complexes pour l'analyse des lois d'echelles. Master des Sciences de la Mati'ere, Ecole Normale Sup´erieure de Lyon Universit´e Claude Bernard Lyon 1
8. Al Jumah, A. (2013). Denoising of an image using discrete stationary wavelet transform and various thresholding techniques. *J Signal Inform Process 4*(2013), 33–41. https://doi.org/10.4236/jsip.2013.41004
9. Demirel, H., & Anbarjafari, G. (2011). Image resolution enhancement by using discrete and stationary wavelet decomposition. *IEEE Transactions on Geoscience and Remote Sensing, 20*(5), 1458–1460.

10. You, N., Han, L., Zhu, D., & Song, W. (2023). Research on image denoising in edge detection based on wavelet transform. *Applied Sciences, 13*, 1837. https://doi.org/10.3390/app13031837
11. Denoise image using deep neural network—MATLAB denoiseImage (mathworks.com) https://www.mathworks.com/help/images/ref/denoiseimage.html
12. Talbi, M., & Bouhlel, M. S. (2018). Wavelets and LPG-PCA for image denoising. *Wavelet Theory Appl.* https://doi.org/10.5772/intechopen.7445
13. Fadhil, M. J., Fayadh, R. A., & Wali, M. K. (2020). Design and implementation a prototype system for fusion image by using SWT-PCA algorithm with FPGA technique. *International Journal of Electrical and Computer Engineering, 10*(1), 757–766.
14. Patil, C., & Mishra, R. (2018). FPGA implementation of dual tree complex wavelet transform in Verilog HDL. *International Journal Advance Management Technology Engineering Science 8*(3)

A MRI Denoising Technique Based on 2D Dual-Tree DWT and Stationary Wavelet Transform 2D

3

3.1 Introduction

The noise presence in the digital image can occur during image acquisition, transmission, and processing phases [1, 2]. Additive Gaussian white noise (AGWN) is a noise sort that frequently occurs in different systems [3]. Therefore, cancelling this noise sort from a noisy image is needed prior to further processing such as texture analysis, feature extraction, and segmentation [4]. Preserving important features of an image, such as edges and textures, is one of the essential issues faced throughout the denoising procedure [5]. Although due to the fact that the noise, texture, and edge are high-frequency components, it is not easy to discriminate them in the denoising scheme, and the denoised image may ineluctably lose some interesting details [6]. A great number of techniques have been proposed in the literature, and among them, we can mention wavelet transforms and non-local means filters [7]. Concerning the thresholding in the wavelet transform domain, the noisy image is decomposed into the low and high-frequency sub-bands. Then, these sub-bands are thresholded. This thresholding is quite effective when applied to the high-frequency sub-bands. Although, it fails in the case where applied to the low frequency sub-band [8]. The DWT was largely utilized. Although, this transform owns three drawbacks, which are shift invariance, aliasing, and lack of poor directionality [9]. Moreover, the DWT is providing the advantages of adaptation and smoothness, as Donoho and Coifman [10] suggested, the DWT exhibits visual artefacts recognized as Gibbs phenomena in the discontinuities [1]. In order to address this problem, the translation invariant denoising technique which is called cycle spinning, has been applied for cancelling such artefacts. Meanwhile, $DT - CWT$ is a method that was introduced in order to solve the main DWT problems. It was combined with non-local means (NLM) [11–13]. Two well-known wavelet transforms, DWT and $DT - CWT$, have been applied sequentially.

© The Author(s), under exclusive license to Springer Nature Switzerland AG 2026

M. Talbi and B. Nasraoui, *Image Denoising Using Mathematical Tools*, Synthesis Lectures on Engineering, Science, and Technology,
https://doi.org/10.1007/978-3-032-11878-3_3

The low frequency sub-band is only denoised utilizing NLM filtering. Since the high-frequency sub-bands contained noises, these are denoised utilizing hard thresholding with cycle spinning. The DWT inverse is applied to the modified sub-bands in order to have the denoised image [1]. Conventional image denoising methods not only eliminate the noise corrupting an image, but also result the loss in this image. They cannot guarantee the clean cancellation of noise information while conserving the true signal of this image. In order to address the aforementioned issues, an image denoising approach combining an improved threshold function and wavelet transform has been introduced in [14]. In contrary to classical threshold functions, the improved threshold function is a continuous function which can avoid the pseudo-Gibbs effect after image denoising and therefore permits to improve the image quality. During the process, the output image of the finite ridge wave transform is firstly combined with the wavelet transform for ameliorating the denoising performance. Then, an improved threshold function is introduced in order to enhance the reconstructed image quality [14]. Our image denoising approach previously introduced in [15, 16] consists firstly in applying an SWT 2D-based denoising method [17] to each of the noisy wavelet coefficients, denoted as Wb$\{j\}\{s\}$,$1 \leq j \leq 2$, $1 \leq s \leq 3$. Those coefficients are obtained by applying the 2D dual-tree DWT to the noisy image. The denoising method [17] based on SWT 2D consists in applying the soft thresholding to each noisy wavelet coefficient obtained from the noisy image decomposition utilizing the SWT 2D. For this decomposition, the Daubechies four, denoted as "db4", is utilized as the mother wavelet, and the level of this decomposition is equal to 5. The novelty of our proposed image denoising technique [15, 16] was lieing on the application of the thresholding technique in the SWT 2D domain, not to the whole noisy image but to each of the coefficients, Wb$\{j\}\{s\}$,$1 \leq \mathrm{j} \leq 2$, $1 \leq \mathrm{s} \leq 3$, for denoising it. Then we obtain denoised coefficients, named Wd$\{j\}\{s\}$,$1 \leq j \leq 2$, $1 \leq s \leq 3$ to them is applied the inverse of the 2D dual-tree DWT for finally having a denoised image, denoted as I_d. The rest of this chapter is organized as follows: In part 2, materials and methods are described. In part 3, the obtained results and discussion are presented. Finally, the conclusion is given at part 4.

Materials and Methods

A SWT 2D-Based Image Denoising Approach

A denoising technique based on the soft thresholding in the SWT 2D domain [17] is applied in our denoising system introduced in [15, 16]. This technique [17] is summarized by the block diagram shown at Fig. 3.1.

According to (Fig. 3.1), the SWT 2D is firstly applied to the noisy image (denoted as I_b) for having noisy stationary wavelet coefficients. The latters are then denoised by applying the soft thresholding. Finally, the inverse of (SWT^{-1} 2D) is applied to the thresholded coefficients (obtained at the second step), for obtaining the denoised image (denoted as

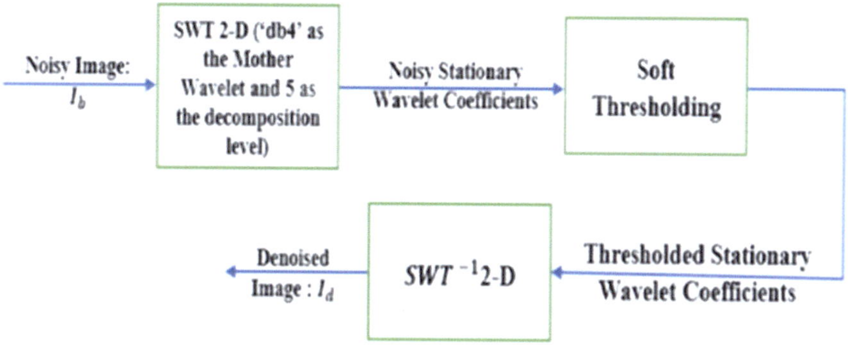

Fig. 3.1 Block diagram of a denoising technique based on the soft thresholding in the SWT 2D domain [17]

I_d). As shown in Fig. 3.1, for applying the SWT 2D, the used mother wavelet is the Daubechies four, denoted as "db4" and the decomposition level is equal to 5.

3.2 The Proposed Image Denoising Technique [15, 16]

In this part, we will detail our denoising approach introduced in [15, 16]. It can be summarized by the block diagram illustrated at Fig. 3.2. As illustrated in Fig. 3.2, the 2D dual-tree DWT is firstly applied to the noisy image (denoised as I_b) for obtaining noisy wavelet coefficients, denoted as Wb{j}{s},$1 \leq j \leq 2$, $1 \leq s \leq 3$ (Wb{1}{1}, Wb{1}{2}, Wb{1}{3}, Wb{2}{1}, Wb{2}{2}, and Wb{2}{3}). Each of those coefficients is then considered as a noisy image and is denoised by applying the denoising approach based on thresholding in the SWT – 2D domain [17] (Fig. 3.1). The denoised wavelet coefficients obtained in this step are denoted as Wd{j}{s},,$1 \leq \mathrm{j} \leq 2$, $1 \leq \mathrm{s} \leq 3$ (Wd{1}{1}, Wd{1}{2}, Wd{1}{3}, Wd{2}{1}, Wd{2}{2}, and Wd{2}{3}). The last step of our denoising technique proposed [15, 16], the inverse of the 2D dual-tree DWT is applied to the coefficients, Wd{j}{s}, $1 \leq j \leq 2$, $1 \leq s \leq 3$ for obtaining the denoised image (denoted as I_d). Also for applying the denoising approach based on thresholding in the SWT 2D domain [17] (Fig. 3.1), the db4 is also utilized as the mother wavelet. Moreover, the decomposition level is chosen to be equal to 5. Those choices were justified in [16]. In a study [15, 16], the proposed image denoising technique was applied to grey-scale images, and in this work, it is applied to both grey-scale and magnetic resonance (MR) images in order to prove its performance.

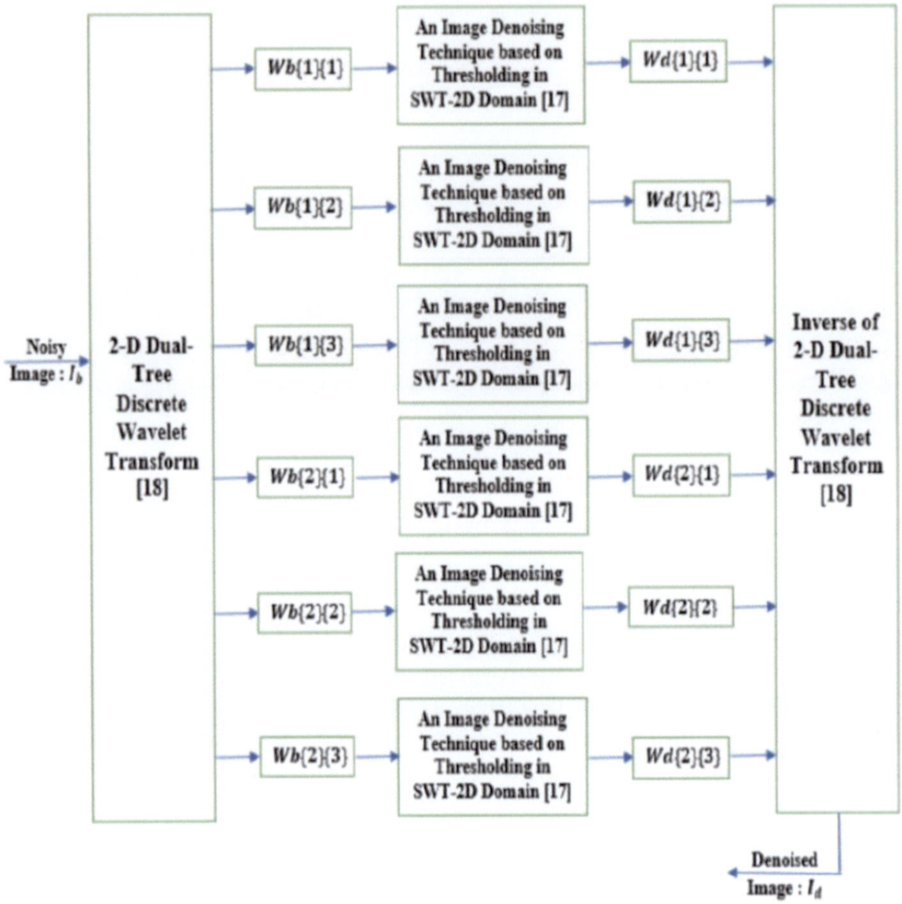

Fig. 3.2 Block diagram of the proposed image denoising approach [15, 16]

3.3 Results and Discussions

In this part, we will present the results obtained by applying our image denoising technique previously proposed in [15, 16] and three other denoising approaches introduced in the literature [17, 18]. These results are in terms of PSNR (peak signal-to-noise ratio) and SSIM (Structural Similarity). These three approaches are the image denoising technique based on thresholding in the SWT 2D domain [17], the image denoising technique based on soft thresholding in the domain of 2D dual-tree DWT [18], and the image denoising technique based on a deep neural network [19]. All the denoising techniques [15–19] were applied for denoising a number of grey-scale images corrupted by an additive Gaussian white noise (AGWN) with diverse values of standard deviation, σ which are 10, 20, 30, and 40, and the obtained results are listed in Table 3.1.

The best results listed in Table 3.1 are highlighted in bold and they show that the image denoising technique introduced in [15, 16] outperforms the three other denoising approaches [17–19], mentioned previously.

Table 3.1 Results in terms of PSNR and SSIM, obtained by applying the four denoising techniques [15–19]

Noisy image	The proposed image denoising Approach [15, 16]	The Image denoising technique based on thresholding in the SWT-2D domain [17]	The image denoising technique based on deep neural network [19]	The image denoising technique based on soft thresholding in the domain of 2-D dual-tree discrete wavelet transform [18]
Noisy st.tif (σ = 10)	PSNR: **34.2517** SSIM: 0.8791	PSNR: 33.7831 SSIM: **0.8962**	PSNR: 28.3030 SSIM: 0.5425	PSNR: 34.1453, SSIM: 0.8675
Noisy st.tif (σ = 20)	PSNR: **30.6126** SSIM: 0.7722	PSNR: 30.1884, SSIM: 0.6950	PSNR: 22.4725, SSIM: 0.2850	PSNR: 30.2840, SSIM: 0.7306
Noisy st.tif (σ = 30)	PSNR: **28.1244** SSIM: 0.6131	PSNR: 24.5706, SSIM: 0.3719	PSNR: 18.9533, SSIM: 0.1789	PSNR: 27.7900, SSIM: 0.5977
Noisy st.tif (σ = 40)	PSNR: **26.4861** SSIM: **0.5138**	PSNR: 20.6093, SSIM: 0.2171	PSNR: 16.4559, SSIM: 0.1248	PSNR: 25.9906, SSIM: 0.4953
Noisy Peppers.tif (σ = 10)	PSNR: **32.8048** SSIM: **0.8250**	PSNR: 32.4622, SSIM: 0.8019	PSNR: 28.6273, SSIM: 0.6497	PSNR: 32.6519, SSIM: 0.8180
Noisy Peppers.tif (σ = 20)	PSNR: **29.6501** SSIM: **0.7233**	PSNR: 29.3674, SSIM: 0.6821	PSNR: 22.4648, SSIM: 0.3569	PSNR: 29.5645, SSIM: 0.7038
Noisy Peppers.tif (σ = 30)	PSNR: **27.8670** SSIM: **0.6406**	PSNR: 27.8481, SSIM: 0.6264	PSNR: 18.8973, SSIM: 0.2208	PSNR: 27.4151, SSIM: 0.5968
Noisy Peppers.tif (σ = 40)	PSNR: **26.3256** SSIM: **0.5462**	PSNR: 26.2747, SSIM: 0.5424	PSNR: 16.3708, SSIM: 0.1497	PSNR: 25.7770, SSIM: 0.5109
Noisy House.tif (σ = 10)	PSNR: **35.0397** SSIM: **0.9203**	PSNR: 34.2284, SSIM: 0.8466	PSNR: 28.4578, SSIM: 0.5576	PSNR: 35.0360, SSIM: 0.9121
Noisy House.tif (σ = 20)	PSNR: **30.7882** SSIM: **0.8462**	PSNR: 30.3082, SSIM: 0.7360	PSNR: 22.4921, SSIM: 0.3090	PSNR: 30.5116, SSIM: 0.7766
Noisy House.tif (σ = 30)	PSNR: **28.1921** SSIM: **0.6586**	PSNR: 27.0232, SSIM: 0.5584	PSNR: 18.9670, SSIM: 0.2049	PSNR: 27.9388, SSIM: 0.6487
Noisy House.tif (σ = 40)	PSNR: **26.2186** SSIM: 0.5368	PSNR: 25.4000, SSIM: 0.4792	PSNR: 16.4148, SSIM: 0.1455	PSNR: 26.0214, SSIM: 0.5416

In the following section, some results are listed for different image denoising examples. Those examples are obtained by applying our denoising technique introduced in [15, 16] (Fig. 3.2). Those results are also in terms of PSNR and SSIM [15, 16]. Those examples are shown in Figs. 3.3, 3.4 and 3.5.

Figures. 3.3, 3.4 and 3.5 and the results (in terms of PSNR and SSIM) obtained by applying the image denoising approach proposed in [15, 16] show the performance of this approach. In fact, this approach [15, 16] allows considerable reduction of the noise and permits to conserve the diverse details of the original images, particularly in cases where the PSNR1 (PSNR before denoising) is higher (Figs. 3.3 and 3.5). Moreover, this introduced technique permits considerable amelioration in the PSNR values and the obtained SSIM values are near 1 (Figs. 3.3 and 3.5).

We also applied our denoising technique introduced [15, 16] to a number of magnetic resonance (MR) images. Figures 3.6, 3.7, 3.8 and 3.9 show four examples of MR image denoising by applying this denoising approach introduced in [15, 16]. Those figures and

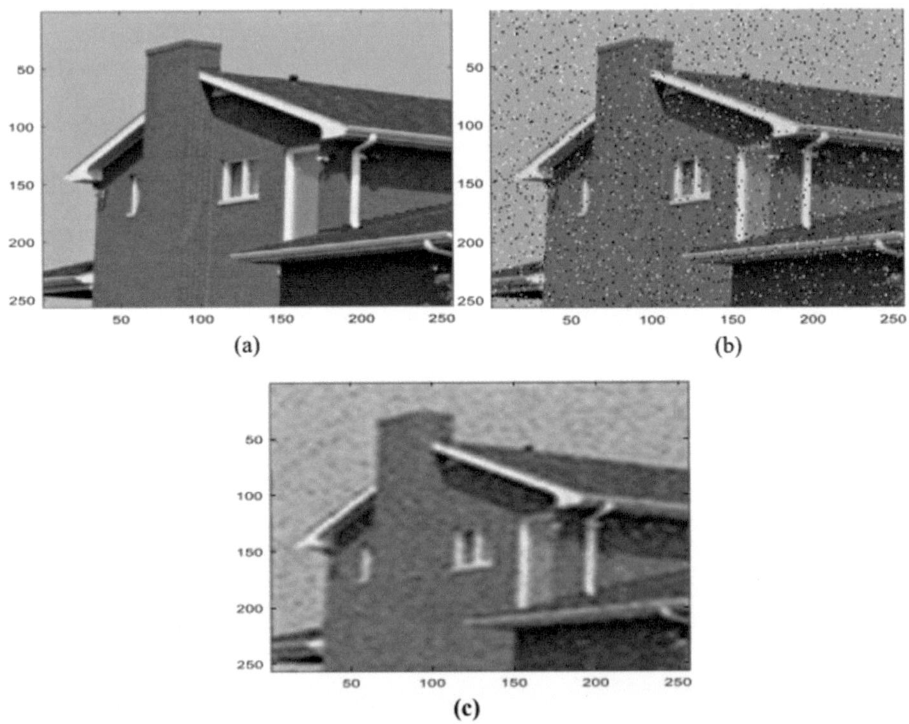

Fig. 3.3 First example of image denoising: **a** Clean image ("house.tif"), **b** noisy image (clean image ("house.tif") corrupted by an additive salt and peppers noise with PSNR = 66.5736 dB, SSIM = 0.9997), **c** denoised image obtained via the application of the approach introduced [15, 16] (PSNR = 74.7058 dB, SSIM = 0.9999)

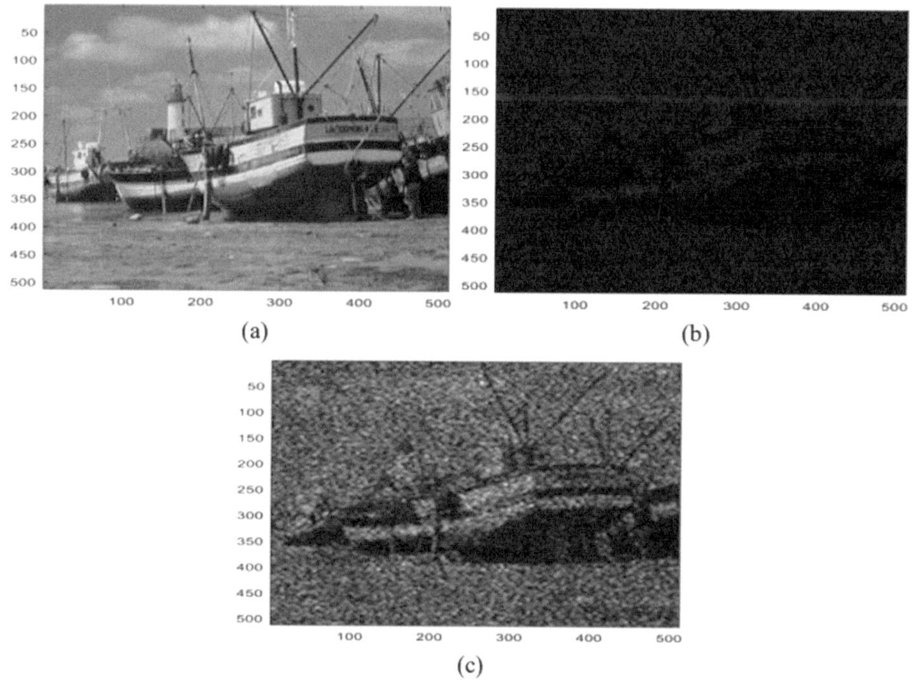

Fig. 3.4 Second example of image denoising: **a** Clean image ("boat.tif"), **b** noisy image (clean image ("boat.tif") corrupted by adding Poisson noise with PSNR = 51.0684 dB, SSIM = 0.9893), **c** denoised image obtained via the application of the image denoising approach introduced [15, 16] (PSNR = 63.6801 dB, SSIM = 0.9980)

the results (in terms of PSNR and SSIM) obtained by applying our previously proposed technique [15, 16] to four MR images degraded by AGWN, show the performance of this technique [15, 16]. In fact, in case where the noise level, σ, is low ($\sigma = 10$ and $\sigma = 20$ as illustrated in Figs. 3.6 and 3.9), this technique permits the elimination of the noise from the noisy image while preserving the diverse details of the original image. Though, in case where $\sigma = 30$, $\sigma = 40$ (as illustrated in Figs. 3.7 and 3.8), this technique permits the cancellation of a great amount of noise while introducing some degradations on the original images.

In this part, we also present a comparative study between the proposed image denoising approach [15, 16] and the non-local means filter [20]. This comparative study was performed by calculating PSNR, SSIM, NMSE (normalized mean square error) and feature similarity (FSIM) and the obtained results are listed in Tables 3.2, 3.3, 3.4, 3.5, 3.6, 3.7, 3.8, 3.9, 3.10, 3.11, 3.12 and 3.13. According to the results listed in these tables, the best results are highlighted in bold and are approximately obtained by our image denoising technique introduced in [15, 16]. Therefore, this technique [15, 16] outperforms the

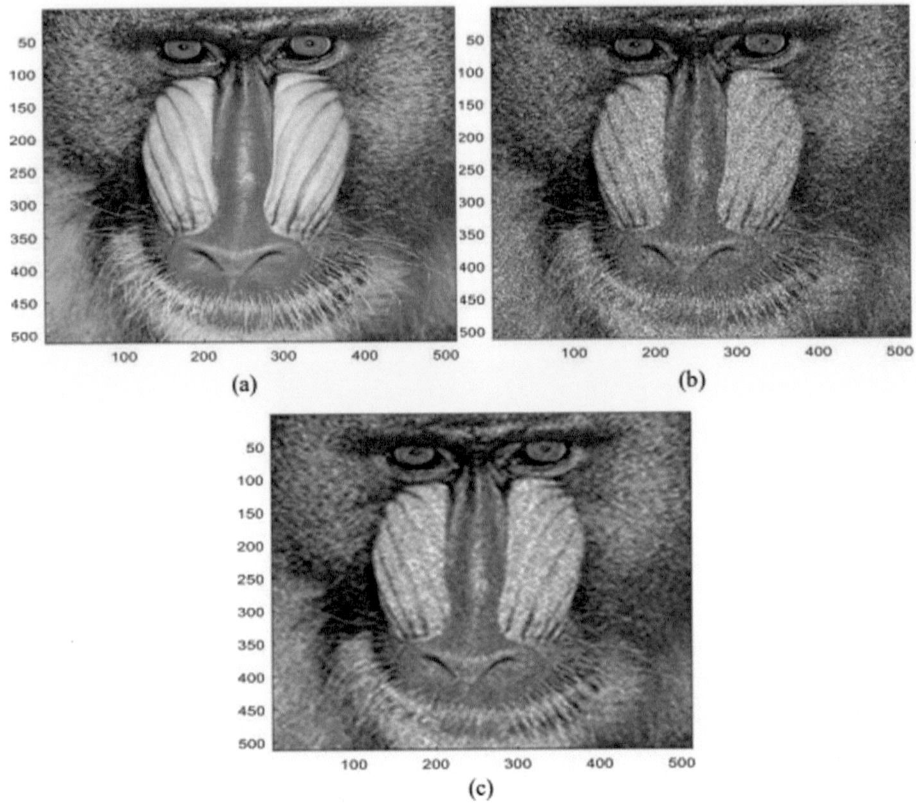

Fig. 3.5 Third example of image denoising: **a** Clean image ("mandrill.tif"), **b** noisy image (clean image ("mandrill.tif") corrupted by adding Poisson noise with PSNR1 = 63.9702 dB, SSIM1 = 0.9994), **c** denoised image obtained via the application of the approach introduced [15, 16] (PSNR = 69.1995 dB, SSIM = 0.9998)

non-local means filter [20]. Figures 3.10, 3.11, 3.12, 3.13, 3.14, 3.15, 3.16, 3.17, 3.18, 3.19, 3.20 and 3.21 show the performance of the image denoising approach introduced in [15, 16]. In fact, when $\sigma = 10$ or $\sigma = 20$, this technique proposed in [15, 16] allows the cancellation of the noise from the noisy images while introducing slight distortions on the diverse details of the original images. Though, when $\sigma = 30$ or $\sigma = 40$, this technique [15, 16] permits to cancel a great amount of the noise while introducing some distortions on the original images.

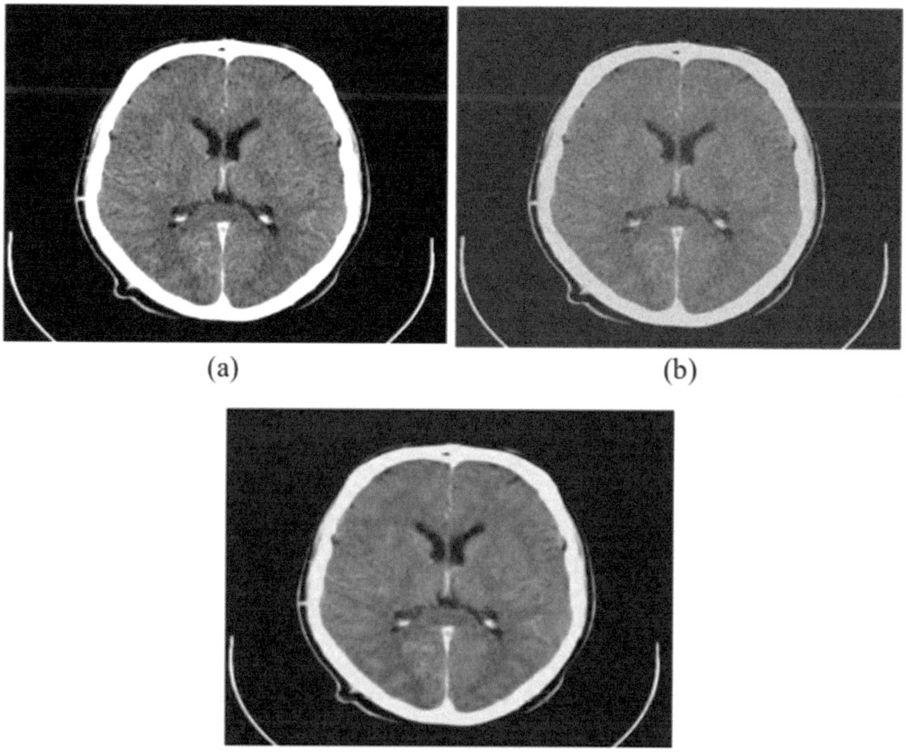

(a) (b) (c)

Fig. 3.6 First example of image denoising: **a** Clean image ("0_fNRjMcFmTjiC3c1K.png"), **b** noisy image (clean image ("0_fNRjMcFmTjiC3c1K.png") degraded by adding a Gaussian white noise with the noise level, $\sigma = 20$, PSNR1 = 22.1376 dB, and SSIM1 = 0.3229, **c** denoised image obtained via the application of the approach introduced [15, 16] (PSNR2 = 28.7066 dB, SSIM2 = 0.6357)

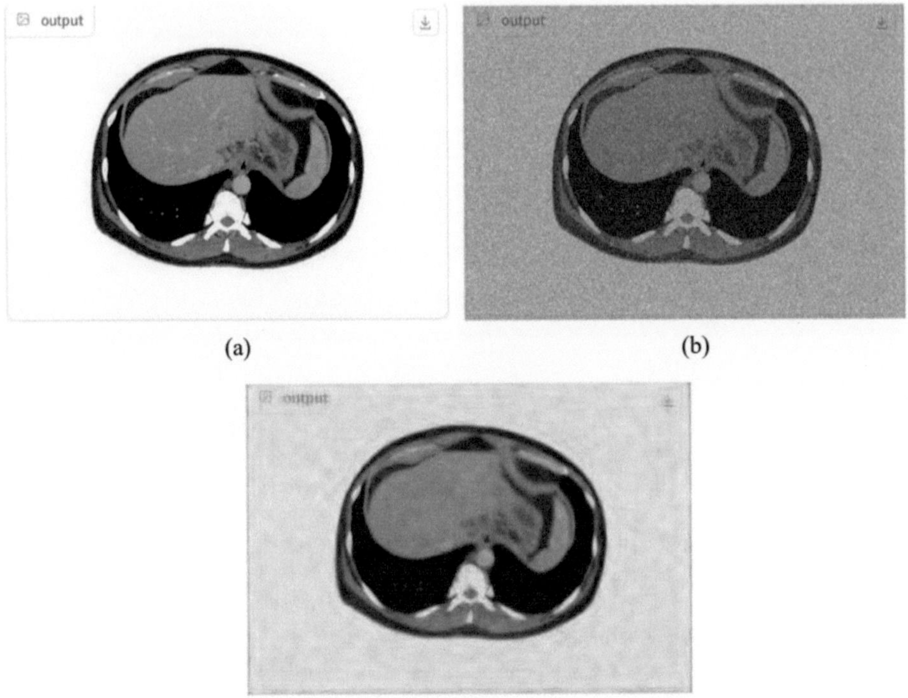

(a) (b) (c)

Fig. 3.7 Second example of image denoising: **a** Clean image ("232,001,710-094d1257-f1fa-479d-98be-725c8673e80b.png"), **b** noisy image (clean image ("232,001,710-094d1257-f1fa-479d-98be-725c8673e80b.png") corrupted by adding a Gaussian white noise with the noise level, $\sigma = 30$, PSNR1 = 18.5782 dB, and SSIM1 = 0.2169, **c** denoised image obtained via the application of the denoising scheme introduced [15, 16] (PSNR2 = 26.7592 dB, SSIM2 = 0.6211)

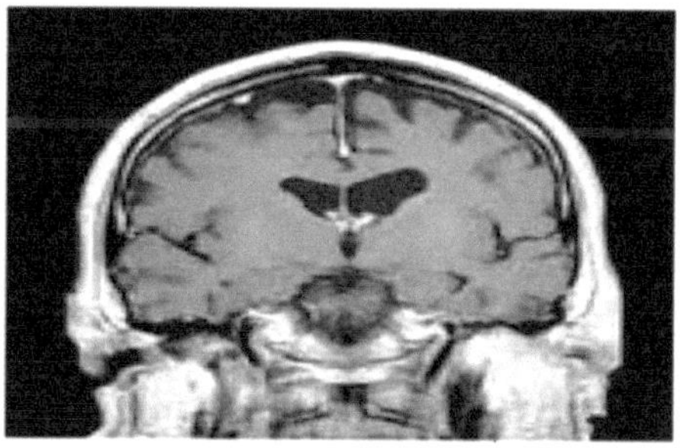

(a)

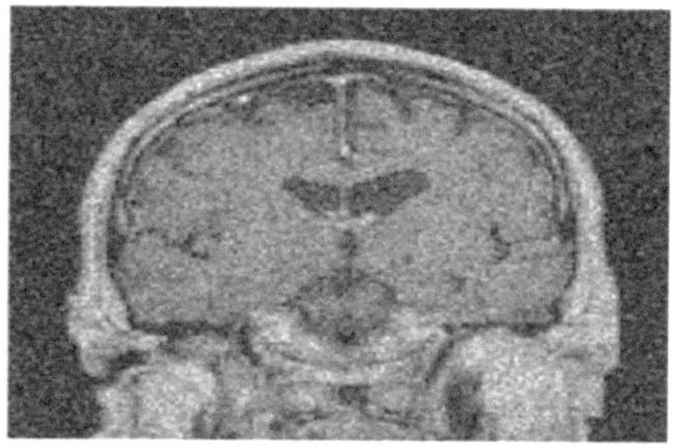

(b)

Fig. 3.8 Third example of image denoising: **a** Clean image ("download.jpg"), **b** noisy image (clean image ("download.jpg") corrupted by adding a Gaussian white noise with the noise level, $\sigma = 40$, PSNR1 = 16.0831dB, and SSIM1 = 0.4106, **c** denoised image obtained via the application of the approach introduced in [15, 16] (PSNR2 = 22.0748 dB, SSIM2 = 0.5760)

Fig. 3.8 (continued)

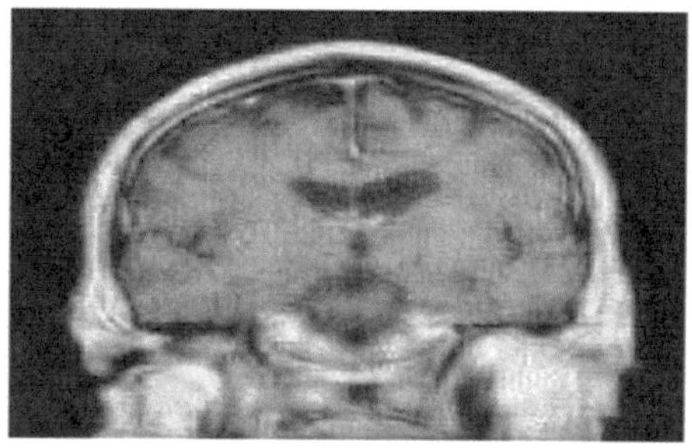

(c)

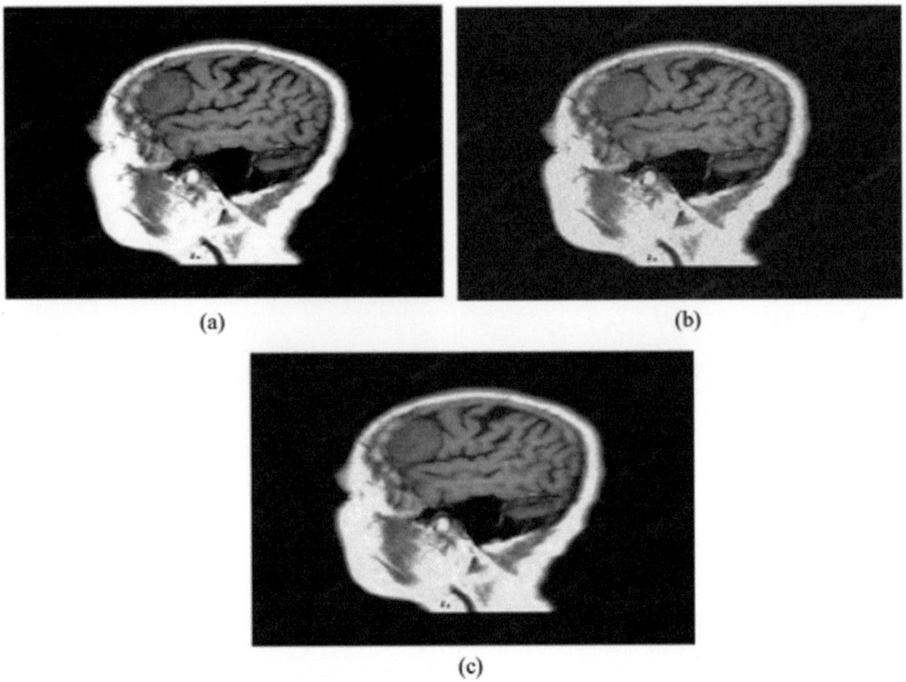

Fig. 3.9 Fourth example of image denoising: **a** Clean image ("800wm.jpg"), **b** noisy image (clean image ("800wm.jpg") corrupted by adding a Gaussian white noise with the noise level, σ = 10, PSNR1 = 28.1511dB, and SSIM1 = 0.5736, **c** denoised image obtained via the application of the approach introduced [15, 16] (PSNR2 = 33.0660 dB, SSIM2 = 0.8152)

Table 3.2 First example of MRI denoising: clean image ("input_MRI.jpg") corrupted by AGWN with standard deviation, σ = 10

Sigma = 10	The proposed technique [15, 16]	Non-local means filter [20]
PSNR	37.8076	36.7189
SSIM	0.9219	0.8640
NMSE	0.0869	0.1375
FSIM	0.9664	0.9523

Table 3.3 Second example of MRI denoising: clean image ("input_MRI.jpg") corrupted by AGWN with standard deviation, $\sigma = 20$

Sigma = 20	The proposed technique [15, 16]	Non-local means filter [20]
PSNR	33.5656	31.4891
SSIM	0.7884	0.6434
NMSE	0.0705	0.1077
FSIM	0.9339	0.8883

Table 3.4 Third example of MRI denoising: clean image ("input_MRI.jpg") corrupted by AGWN with standard deviation, $\sigma = 30$

Sigma = 30	The proposed technique [15, 16]	Non-local means filter [20]
PSNR	30.7345	28.2278
SSIM	0.6395	0.4539
NMSE	0.0725	0.1227
FSIM	0.8997	0.8173

Table 3.5 Fourth example of MRI denoising: clean image ("input_MRI.jpg") corrupted by AGWN with standard deviation, $\sigma = 40$

Sigma = 40	The proposed technique [15, 16]	Non-local means filter [20]
PSNR	27.9419	25.8125
SSIM	0.4527	0.3196
NMSE	0.0856	0.1522
FSIM	0.8551	0.7521

Table 3.6 Fifth example of MRI denoising: clean image ("source18B.tif") corrupted by AGWN with standard deviation, $\sigma = 10$

Sigma = 10	The proposed technique [15, 16]	Non-local means filter [20]
PSNR	34.6734	34.9287
SSIM	0.8694	0.8189
NMSE	0.2019	0.3301
FSIM	0.9649	0.9613

Table 3.7 Sixth example of MRI denoising: clean image ("source18B.tif") corrupted by AGWN with standard deviation, σ = 20

Sigma = 20	The proposed technique [15, 16]	Non-local means filter [20]
PSNR	30.507	30.3047
SSIM	0.6713	0.5932
NMSE	0.1823	0.228
FSIM	0.9213	0.9071

Table 3.8 Seventh example of MRI denoising: clean image ("source18B.tif") corrupted by an AGWN with standard deviation, σ = 30

Sigma = 30	The proposed technique [15, 16]	Non-local means filter [20]
PSNR	27.2755	27.2694
SSIM	0.4620	0.4358
NMSE	0.1649	0.1575
FSIM	0.8701	0.8481

Table 3.9 Eighth example of MRI denoising: clean image ("source18B.tif") corrupted by AGWN with standard deviation, σ = 40

Sigma = 40	The proposed technique	Non-local means filter [20]
PSNR	25.2564	25.0927
SSIM	0.3551	0.3349
NMSE	0.1787	0.1822
FSIM	0.8246	0.7899

Table 3.10 Ninth example of MRI denoising: clean image ("med256B.jpg") corrupted by AGWN with standard deviation, σ = 10

Sigma = 10	The proposed technique	Non-local means filter [20]
PSNR	34.1486	34.9727
SSIM	0.9026	0.8981
NMSE	0.2179	0.191
FSIM	0.9542	0.9579

Table 3.11 Tenth example of MRI denoising: clean image ("med256B.jpg") corrupted by AGWN with standard deviation, $\sigma = 20$

Sigma = 20	The proposed technique	Non-local means filter [20]
PSNR	30.7208	30.4004
SSIM	0.7978	0.7559
NMSE	0.1711	0.1762
FSIM	0.9159	0.9152

Table 3.12 Eleventh example of MRI denoising: clean image ("med256B.jpg") corrupted by AGWN with standard deviation, $\sigma = 30$

Sigma = 30	The proposed technique	Non-local means filter [20]
PSNR	28.0427	27.1948
SSIM	0.6728	0.6146
NMSE	0.1752	0.1941
FSIM	0.8782	0.8664

Table 3.13 Twelfth example of MRI denoising: clean image ("med256B.jpg") corrupted by an additive Gaussian white noise (AGWN) with standard deviation, $\sigma = 40$

Sigma = 40	The proposed technique	Non-local means filter [20]
PSNR	26.467	25.0481
SSIM	0.6036	0.5121
NMSE	0.1505	0.2163
FSIM	0.845	0.8203

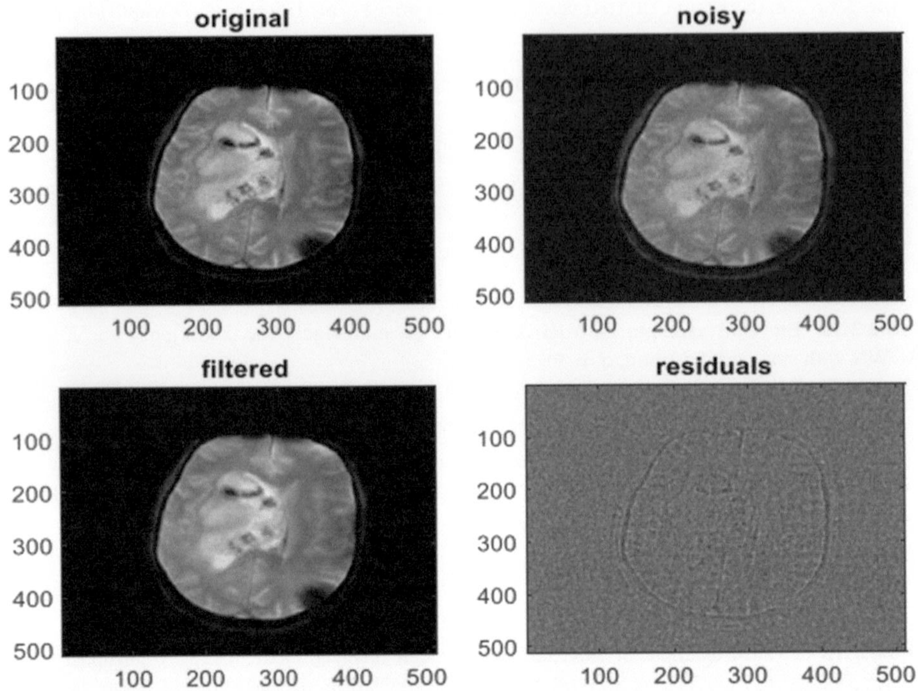

Fig. 3.10 First example of MRI Denoising by applying the proposed image denoising technique

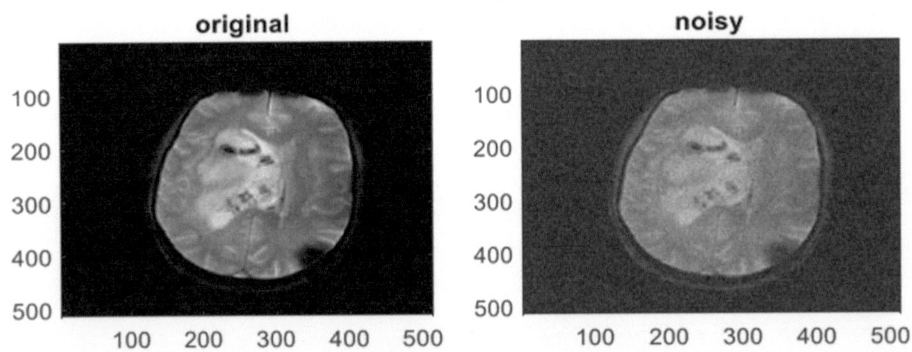

Fig. 3.11 Second example of MRI denoising by applying the proposed image denoising approach

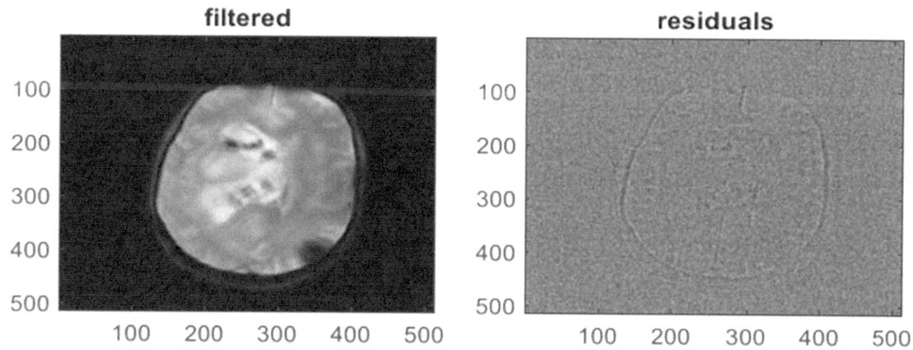

Fig. 3.11 (continued)

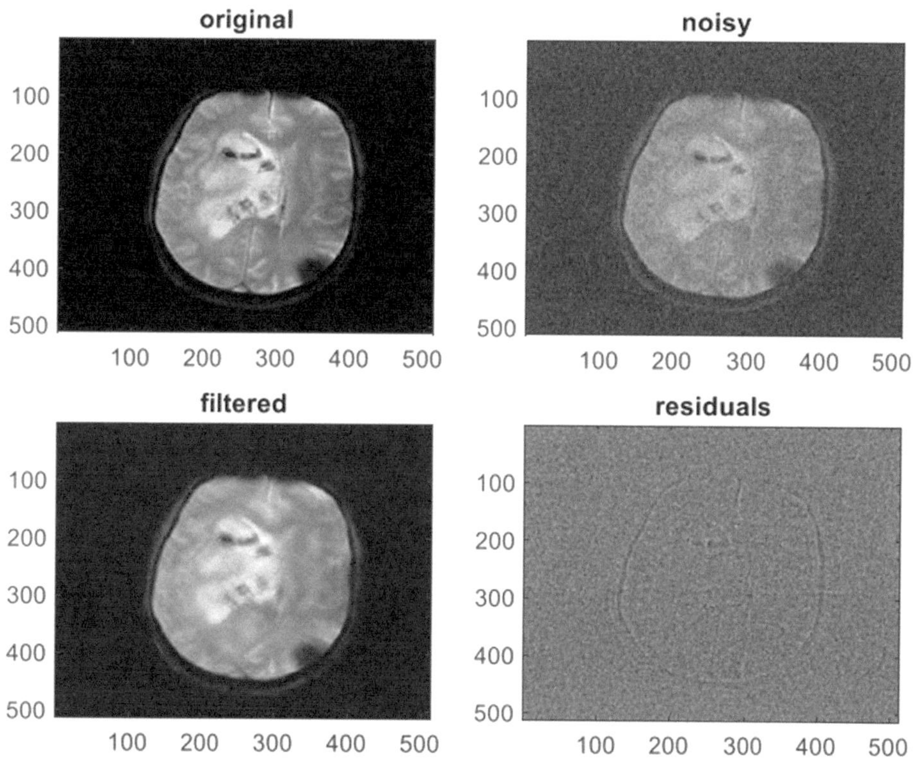

Fig. 3.12 Third example of MRI denoising by applying the proposed image denoising technique

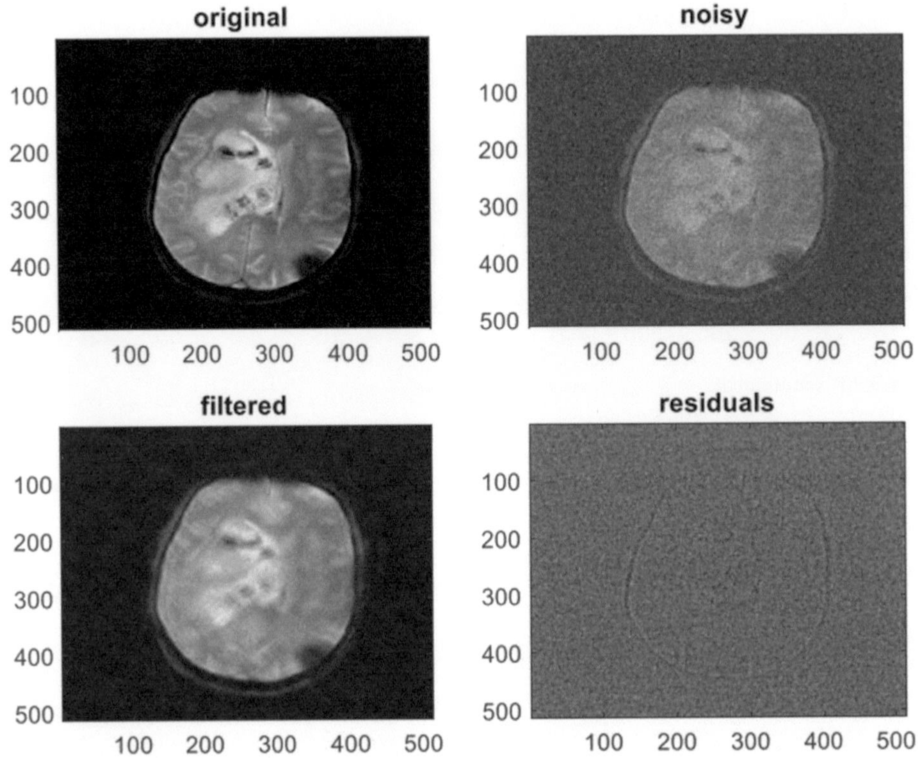

Fig. 3.13 Fourth example of MRI denoising by applying the proposed image denoising technique

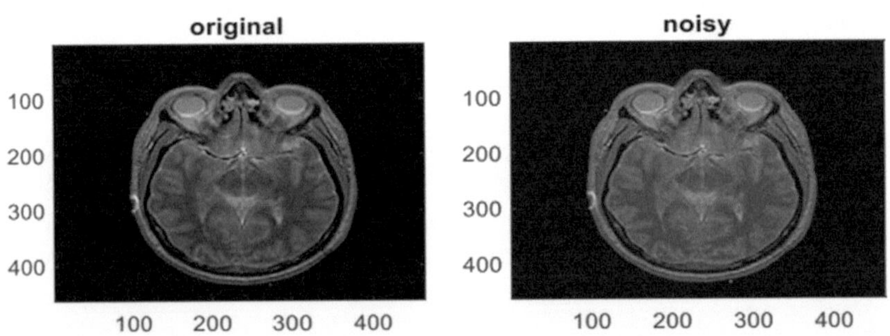

Fig. 3.14 Fifth example of MRI denoising by applying the proposed image denoising technique

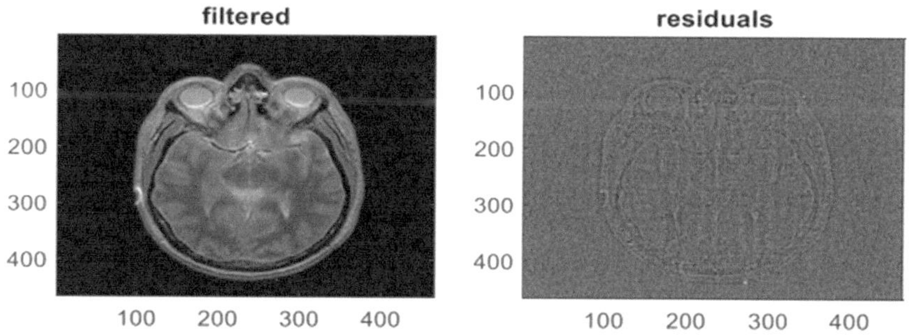

Fig. 3.14 (continued)

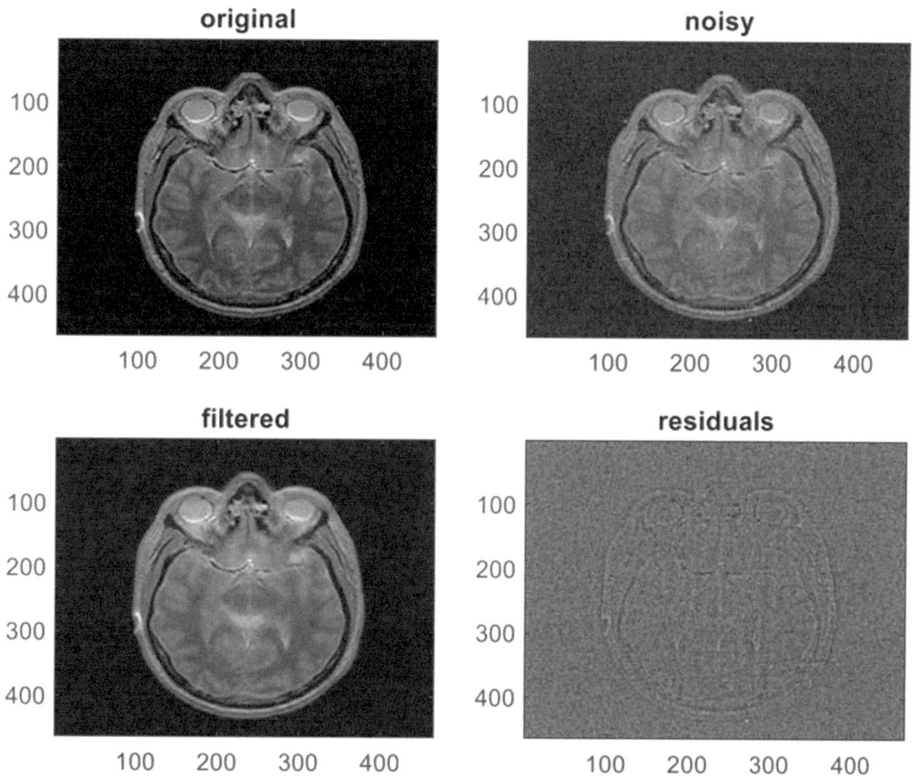

Fig. 3.15 Sixth example of MRI denoising by applying the proposed image denoising technique

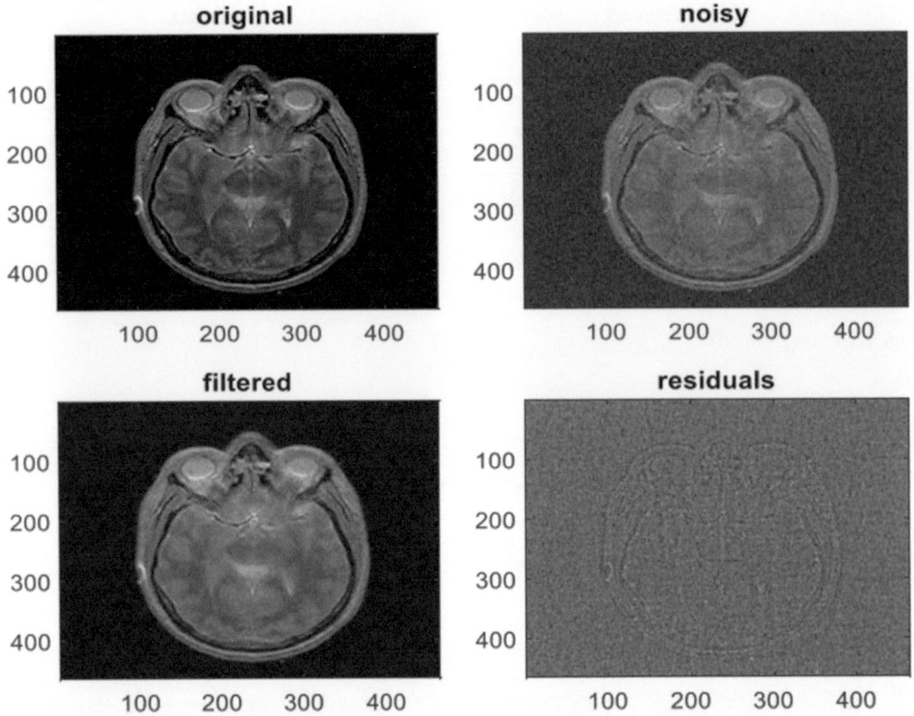

Fig. 3.16 Seventh example of MRI denoising by applying the proposed image denoising technique

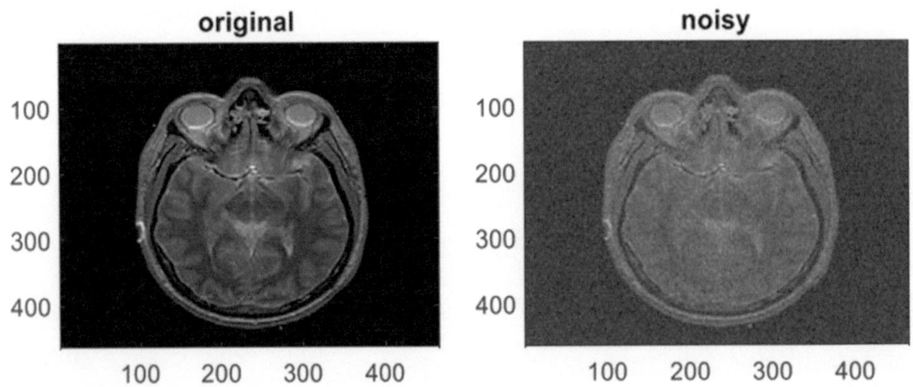

Fig. 3.17 Eighth example of MRI denoising by applying the proposed image denoising technique

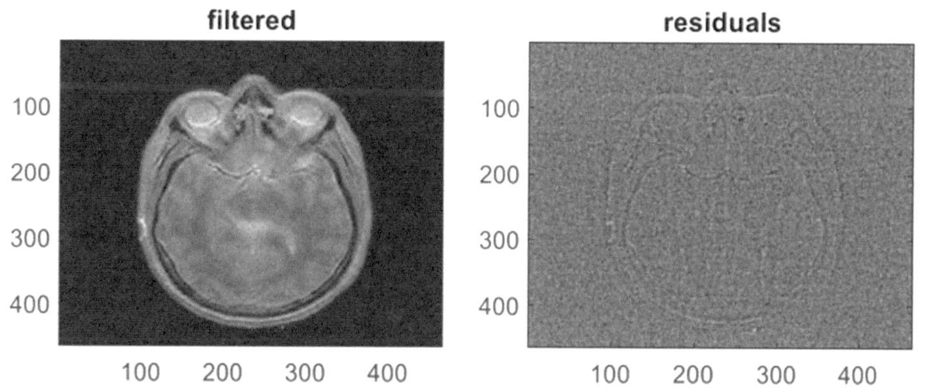

Fig. 3.17 (continued)

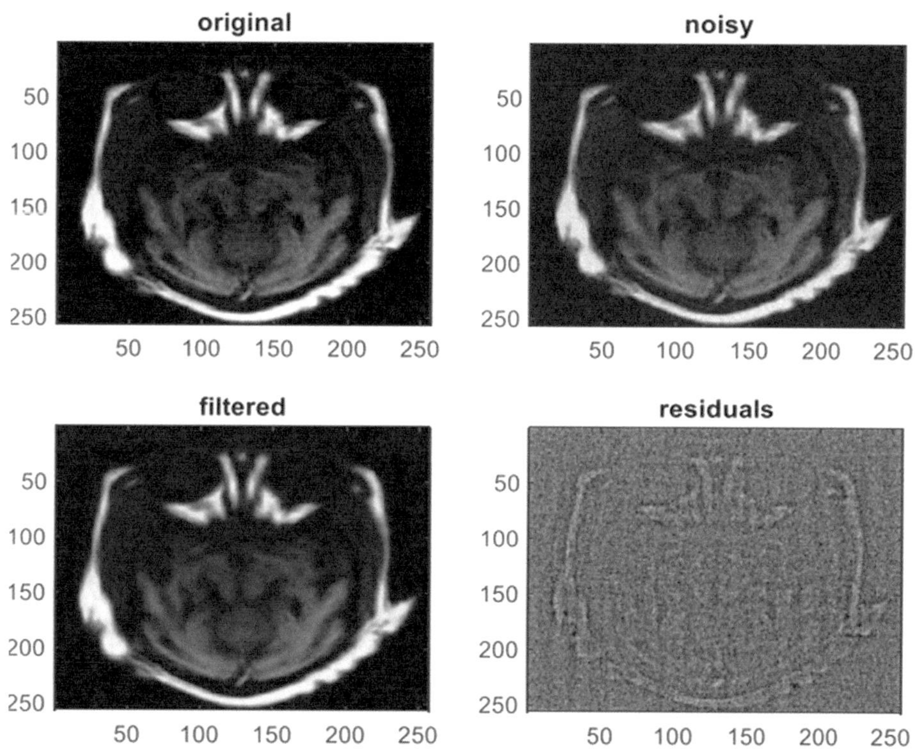

Fig. 3.18 Ninth example of MRI denoising by applying the proposed image denoising technique

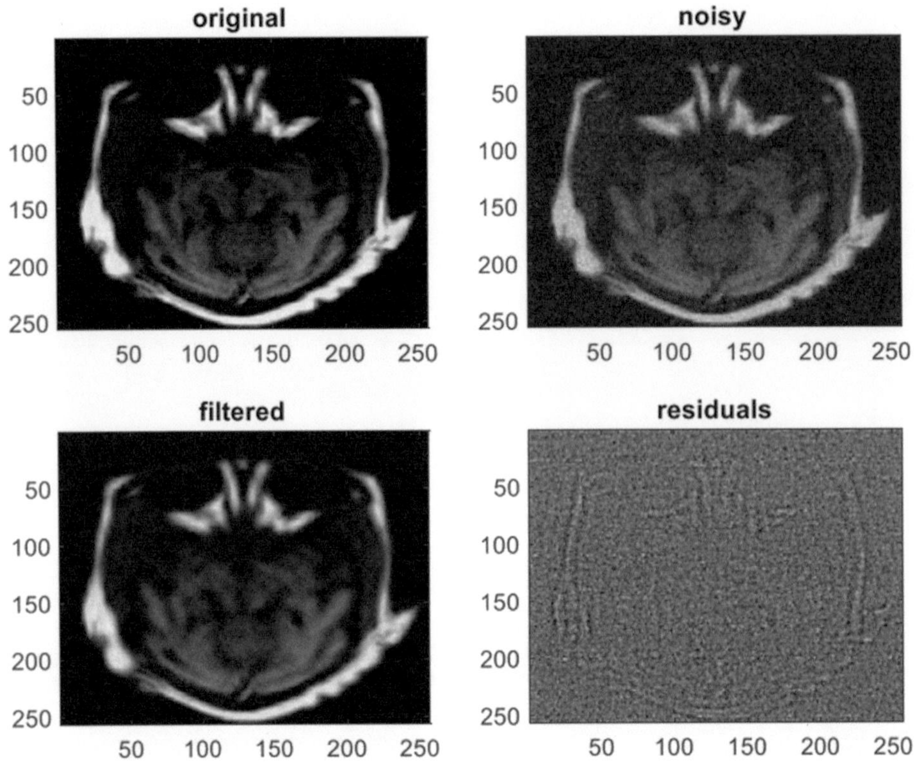

Fig. 3.19 Tenth example of MRI denoising by applying the proposed image denoising technique

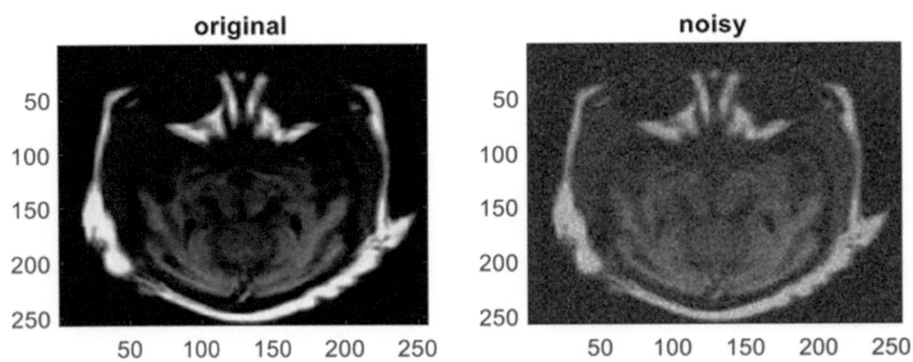

Fig. 3.20 Eleventh example of MRI denoising by applying the proposed image denoising technique

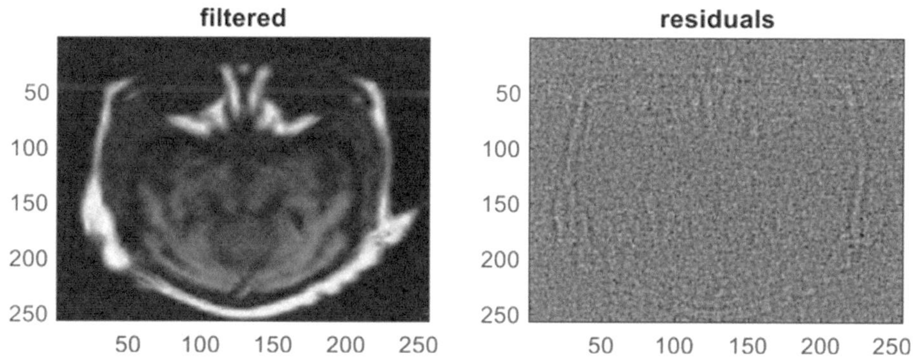

Fig. 3.20 (continued)

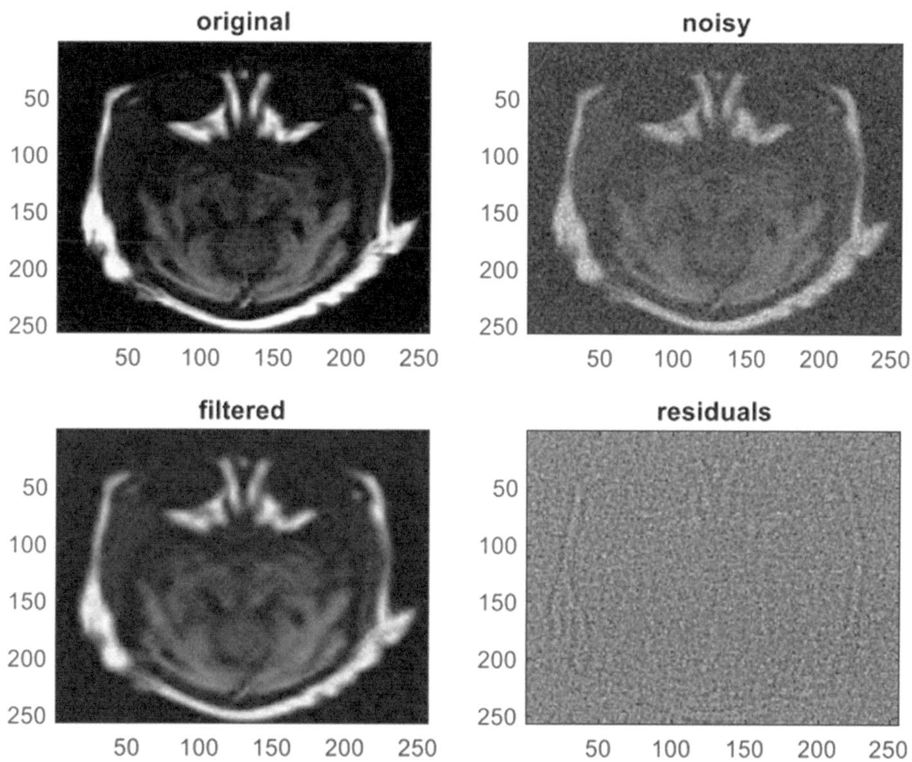

Fig. 3.21 Twelfth example of MRI denoising by applying the proposed image denoising technique

3.4 Conclusion

In this chapter we detail a comparative study of a number of image denoising techniques introduced in literature. These techniques include our proposed image denoising approach based on the SWT 2D and the 2D dual-tree DWT. This approach consists firstly in applying the 2D dual-tree DWT to the noisy image to obtain noisy wavelet coefficients. Secondly, the denoising of each of those coefficients is performed by applying an SWT 2D-based denoising technique. The denoised image is finally obtained by applying the inverse of the 2D dual-tree DWT to the denoised coefficients obtained in the second step. The performance of this technique is proven by its comparison to three other techniques, which are the soft thresholding in the SWT 2D domain, the soft thresholding in the 2D dual-tree DWT domain and the denoising technique based on a deep neural network. All those approaches, including the denoising technique proposed in [15, 16], are applied to a number of noisy images, and the obtained results are in terms of PSNR and SSIM. Those results show that this technique [15, 16] outperforms the three other image denoising techniques [17–19]. We have also applied this proposed approach [15, 16] to a number of noisy MR images, and the results are obtained by computing the PSNR and SSIM. Those results prove the performance of our technique introduced in [15, 16]. In fact, in cases where the noise level $\sigma = 10$ and $\sigma = 20$, this technique permits the elimination of the noise from the noisy images and preserves the divers details of the original images. However, when $\sigma = 30$ or $\sigma = 40$, this technique allows the elimination of a great amount of noise and introduces some distortions on the original images. We also made a comparative study between the approach introduced in [15, 16] and the non-local means filter [20]. This study was performed by calculating PSNR, SSIM, normalized mean square error (NMSE) and feature similarity (FSIM) and the obtained results showed that this proposed approach [15, 16] outperforms the non-local means filter [20].

Abbreviations List

AGWN	Additive Gaussian white noise
db4	Daubechies four
DWT	Discrete wavelet transform
DT-CWT	Dual-tree complex wavelet transform
FSIM	Feature similarity index
I_b	Noisy image
I_d	Denoised image
MRIs	Magnetic resonance images
NLM	Non-local means
NMSE	Normalized mean square error
σ	Standard deviation (the noise level)
PSNR	Peak signal-to-noise ratio
SSIM	Structural similarity index

$\boldsymbol{SWT}2-\boldsymbol{D}$	Two-dimensional stationary wavelet transform
$\boldsymbol{SWT}^{-1}$ **2D**	Inverse two-dimensional stationary wavelet transform
$\boldsymbol{Wb}\{j\}\{s\}, 1 \le j \le 2, 1 \le s \le 3$	Noisy wavelet coefficients
$\boldsymbol{Wd}\{j\}\{s\}, 1 \le j \le 2, 1 \le s \le 3$	Denoised wavelet coefficients

References

1. Karyono, G., Ahmad, A., Asmai, S. A. (2023). Image denoising using wavelet cycle spinning and non-local means filter. *International Journal of Advance Computer Science and Application, 14*(3). https://doi.org/10.14569/IJACSA.2023.0140356
2. Nik, M. M. P., Se, S. V. H. (2018). A review paper: study of various types of noises in digital images. *International Journal of Engineering Trends and Technology, 57*(1), 40–3. https://doi.org/10.14445/22315381/IJETT-V57P208
3. Lin, C. Y., Wu, M., Bloom, J. A., Cox, I. J., Miller, M. (2001). Rotation, scale, and translation resilient public watermarking for images. *IEEE Trans Image Process, 10*(5), 767–782. Li, Y., Liu, C., You, X., & Liu, J. (2022). A single-image noise estimation algorithm based on pixel-level low-rank low-texture patch and principal component analysis. *Sensors (Basel)*. https://doi.org/10.3390/s22228899
4. Rani, S., Chabrra, Y., Malik, K. (2022). An improved denoising algorithm for removing noise in color images. *Engineering, Technology & Applied Science Research, 12*(3), 8738–44. https://doi.org/10.48084/etasr.4952
5. Al-Shamasneh, A. R., Ibrahim, R. W. (2023). Image denoising based on quantum calculus of local fractional entropy. *Symmetry (Basel), 15*(2), 396. https://doi.org/10.3390/sym15020396
6. Fan, L., Zhang, F., Fan, H., Zhang, C. (2019). Brief review of image denoising techniques. *Visual Computing for Industry, Biomedicine, and Art, 2*(1), 7. https://doi.org/10.1186/s42492-019-0016-7. PMID: 32240414
7. Hu, H., Li, B., Liu, Q. (2016). Removing mixture of gaussian and impulse noise by patch-based weighted means. *Journal Science Computer, 67*(1), 103–29. https://doi.org/10.1007/s10915-015-0073-9
8. Singh, P., Pradhan, G. (2017). Denoising of ECG signal by non-local estimation of approximation coefficients in DWT. *Biocybernetics and Biomedical Engeering, 37*(3), 599–610. https://doi.org/10.1016/j.bbe.2017.06.001
9. Mohan, C. R., Chouhan, K., Rout, R. K. et al. (2022). Improved procedure for multi-focus images using image fusion with qshiftN DTCWT and MPCA in laplacian pyramid domain. *Applied Science (Basel), 12*(19), 9495. https://doi.org/10.3390/app12199495
10. Coifman, R. R., Donoho, D. L. (1995). Translation-invariant de-noising. In *Lect Notes Stat Wavelets Stat. London: Springer Nature Link 1995*, pp 125–50. https://doi.org/10.1007/978-1-4612-2544-7_9
11. Shamsi, Z. H., Kim, D. (2011). Multiscale hybrid Non-local means filtering using modified similarity measure. In *Mathematical Problems in Engineering*, pp 1–7. https://doi.org/10.48550/arXiv.1304.0839
12. Rabbouch, H., Saâdaoui, F. (2018). A wavelet-assisted subband denoising for tomographic image reconstruction. *Journal of Visual Communication Image Represent, 55*, 115–30. https://doi.org/10.1016/j.jvcir.2018.05.004

13. Raj, N. P., Venkateswarlu, T. Denoising of MR images using adaptive multiresolution subband mixing. https://doi.org/10.1109/ICCIC.2013.6724247
14. Li, B., Cong, Y., Mo, H. (2024). Image denoising method integrating ridgelet transform and improved wavelet threshold. *PLoS One, 19*(9), e0306706. https://doi.org/10.1371/journal.pone.0306706. PMID: 39240820
15. Mourad, T., Riadh, B. (2023). Perspective chapter: new image denoising approach based on SWT and 2-D dual-tree discrete wavelet transform. In *London, UK: Intechopen 2023*. https://doi.org/10.5772/intechopen.112103
16. Mourad, T., Riadh, B., Brahim, N. (2024). A novel method of image denoising based on 2D dual-tree DWT and SWT. *International Journal of Wavelets, Multiresolution and Information Processing, 22*(4), 2450009. https://doi.org/10.1142/S0219691324500097
17. Generate MATLAB code for 2-D stationary wavelet denoising. (2024). Available from: https://www.mathworks.com/help/wavelet/ug/generating-matlab-codefor-2-d-stationary-wavelet-denoising.html
18. Wavelet software at Brooklyn Poly. (2019). Available from: https://eeweb.engineering.nyu.edu/iselesni/WaveletSoftware/
19. Denoise image using deep neural network. (2020). Available from: https://www.mathworks.com/help/images/ref/denoiseimage.html
20. Non-local means filter. (2025). Available from: https://www.mathworks.com/matlabcentral/fileexchange/13176-non-local-means-filter

Index

© The Editor(s) (if applicable) and The Author(s), under exclusive license to Springer Nature Switzerland AG 2026

M. Talbi and B. Nasraoui, *Image Denoising Using Mathematical Tools*, Synthesis Lectures on Engineering, Science, and Technology,
https://doi.org/10.1007/978-3-032-11878-3

MIX
Papier aus verantwortungsvollen Quellen
Paper from responsible sources
FSC® C105338
FSC
www.fsc.org

If you have any concerns about our products, you can contact us on
ProductSafety@springernature.com

In case Publisher is established outside the EU, the EU authorized representative is:
Springer Nature Customer Service Center GmbH
Europaplatz 3, 69115 Heidelberg, Germany

Printed by Libri Plureos GmbH
in Hamburg, Germany